Job Hazard Analysis

George Swartz

Government Institutes
An imprint of
The Scarecrow Press, Inc.
Lanham, Maryland • Toronto • Oxford

**Government
Institutes**

Published in the United States of America
by Government Institutes, an imprint of The Scarecrow Press, Inc.
A wholly owned subsidary of
The Rowman & Littlefield Publishing Group, Inc.
4501 Forbes Boulevard, Suite 200
Lanham, Maryland 20706
http://www.govinstpress.com/

Estover Road
Plymouth PL6 7PY
United Kingdom

British Library Cataloguing in Publication Information Available

Library of Congress Cataloging-in-Publication Data

Product Code #818

Swartz, George.
    Job hazard analysis: a guide to identifying risk in the workplace / George Swartz.
      p. cm.
    Includes index.
    ISBN 13: 978-0-86587-818-1 (pbk. : alk. paper)
    ISBN 10: 0-86587-818-8 (pbk. : alk. paper)
    1. Job analysis. 2. Hazardous occupations. 3. Industrial Safety. 4. Human-machine systems. I. Title.

HF5549.5.J6 S93 2001
658.3'82—dc21

2001033046

Dedication

This book is dedicated to all of the workers who have lost their lives as a result of industrial injuries and illnesses.

Summary Contents

Contents

List of Figures

List of Tables

Preface

Job hazard analysis (JHA) is the most important tool available to management to help eliminate job hazards and reduce injuries and incidents in the workplace. Also, JHA improves productivity by identifying errors in the production process that can decrease efficiency. Completed JHAs can be used for new employee orientation, for refresher training, for transferred workers, and for aiding injury investigations. Developing written safety procedures is one of the best means of achieving a safer workplace, but management must ensure that these procedures are actually being followed by the workers. In turn, supervisors learn more about the work processes in their department by analyzing the workers who perform them.

Perhaps most importantly, JHA allows workers to become involved in the process and to share their job knowledge. Many programs do not allow for feedback and participation from workers. It is not unusual to hear workers comment that "no one ever asks me for my opinion around here." What makes JHA special is the collaboration that takes place between the supervisor and worker (s) to arrive at the safest method of completing a task or job. Much can be gained by allowing workers to assist in injury investigations, incident reporting, facility inspections, and JHA. A worker is more likely to perform safely if he/she helps to contribute to the development of the safety rules. Without the collaboration of the worker and supervisor, the JHA program will not be as effective.

Job hazard analysis has been around for many years, but in the past an analysis was much narrower than it is today. For one thing, many organizations analyzed only the most hazardous jobs, and for whatever reason; there was not a focus on all jobs. Even if the most dangerous jobs were completed first, jobs of less priority were not focused on. Today, all jobs should be observed for a JHA. Now, ergonomics is one of several factors that the supervisor focuses on while observing and discussing the job with the worker. Many ergonomic issues can be corrected once they have been discovered. JHA can uncover and identify many different kinds of ergonomic problems.

The principles of JHA are not limited to the manufacturing process. These principles can be used in construction, the service industries, in retail establishments, and at schools for instructing students. Even though the graphics in this book are a part of the manufacturing, warehousing, and distribution system, the principles can be utilized just about anywhere. JHA could be used to develop safe procedures if someone were adding a new room to their house or simply shoveling the snow in their driveway.

It is not possible for a book, program, or regulation to provide a safety practice for every contingency that could take place in the workplace. Employers should understand and apply state and federal regulations to the workplace because these regulations safeguard against many potential hazards. Applicable standards are referenced in this book as a way to provide regulatory guidance and identify some significant programs.

To illustrate the JHA process, as well as the types of injuries that occur in the workplace, this book contains many examples of incidents and injuries that have occurred to workers while performing their jobs. In almost every case, the description of the event is real and was personally witnessed or investigated by the author. The examples that have been chosen are common to many work environments, so the analyses provided should be beneficial to the reader.

The focus of this book is to provide guidelines on how to identify potential hazards in the workplace by carefully observing the employee at work and examining the workplace. To accomplish this task, the book provides detailed discussion of how to conduct job observations and how to recognize the sources of injuries. The book includes sample JHA forms and directions for properly recording the basic steps, hazards, and analysis of a job. Completed JHAs for common jobs are also included. These should guide the reader in completing their own JHAs.

The photographs in this book of each job step will aid in understanding the JHA process by providing a step-by-step view of the job tasks. Photographing or videotaping job tasks is very useful. It can reveal much more than the eye can take in during a brief observation. Also, since JHAs are an excellent resource for employee training, photographs and videos of job tasks are effective training resources.

Job hazard analysis is a highly effective alternative to compliance-based safety programs and is an integral part of any effective safety and health program. Additional information and resources for developing effective safety and health programs, including relevant safety and health statistics, are provided in the appendices.

For those readers that are not sure if JHA is for them and need more convincing information to demonstrate that this program is worth the time and effort for your organization, consider the following example provided by Stan F. Spence in a 1978 issue of *Locomotive* magazine: Some years ago in an organic chemical plant a worker was killed when he applied 125-psig steam to clean a gauge glass. Through an investigation of the fatality, the supervisor found that, with the product involved, the gauge glass had to be cleaned regularly because the material was quite viscous. Originally the gauge glass was cleaned simply by the application of low-pressure air. After a while, workers discovered that high-pressure air—on the order of 50 psig or so—did the job better. Sometime later, it was found by certain operators that 90-psig steam did a better job than 50 psig. When the employee was killed, he was using 125-psig steam on the gauge.

The gauge was attached to a vessel that was not designed as a pressure vessel. It was an atmospheric vessel, normally vented to the atmosphere. At the time of the fatality, the vent lines had clogged and the vessel had applied to it the entire line pressure of the steam, which was applied to the gauge glass. The operators were described as living with a "normal malfunction."

A proper JHA of this job could have prevented this injury. "No process was ever designed to undergo a continued malfunction and no continued malfunction should ever be tolerated. Job safety analysis brings these small elements which cause large accidents with extremely serious results to light—and subjects them to the necessary corrective action...It is the authors experience, (Mr. Spence), almost every accident of severe consequence—including fires, explosions, mechanical accidents, and others—could have been prevented had there been a job safety analysis made by first line supervision and the necessary action taken. It is a serious reflection on our inertia that an accident, producing in some cases fatalities, has to occur before a given plant systematically undertakes job safety analysis by first line supervision." Mr. Spence's experience with JSA dates to 1945.

About the Author

George Swartz has been involved in the safety profession for 30 years. He has extensive safety and health experience in manufacturing, warehousing and distribution, automotive repair shops, and heavy construction. His safety experience includes working for Pittsburgh Bridge and Iron as a corporate director of safety, for Envirotech Corporation as a division safety officer and for Midas International as a corporate director of safety and health.

His areas of expertise include: safety auditing, job hazard analysis, warehouse and distribution safety, auto repair safety, forklift safety, asbestos control while servicing brakes, and general plant safety. He has lectured and has provided training at major professional safety conferences, safety councils, corporations, universities, for OSHA compliance officers, and for classes internationally.

Mr. Swartz currently serves on the board of directors for the National Safety Council, is Vice President of the Industrial Division of the National Safety Council and is on the board of advisors for Safety Director.Com. He is also the senior editor for *Compliance* magazine. He has authored articles in *Professional Safety, Safety and Health, Today's Supervisor, Compliance, Safety and Hygiene New*s, and numerous other safety publications. He has been interviewed for a number of articles in industrial trade magazines. He has also authored the following books: *Forklift Safety, Warehouse Safety,* and *Auto Repair Shop Safety.* He was the chief editor of *Safety Culture and Effective Safety Management.*

Mr. Swartz has served as a lecturer and instructor at Northern Illinois University in De Kalb, IL, provided testimony during OSHA hearings on the asbestos standard, and provided assistance to the Voluntary Protection Program Participants Association with articles and training programs at their regional and national conferences.

He has been honored with awards and recognized as a Fellow by the American Society of Safety Engineers, as Safety Professional of the Year by the management division of the American Society of Safety Engineers, with the Distinguished Service to Safety Award by The National Safety Council, for service as chairman for the industrial division of the National Safety Council and as an Outstanding Volunteer. He received the Charles V.Culbertson award from the American Society of Safety Engineers.

Mr. Swartz has been a certified safety professional for more than 20 years. He is currently a safety consultant and lives in Davenport, FL. He has the following degrees: a BA from the University of Pittsburgh in social sciences/education, an MS from Northern Illinois University in safety, a Certificate of Advanced Study in Safety with a thesis specialty in auditing from Northern Illinois University and an MS in Managerial Communications from Northwestern University in Evanston, IL

Acknowledgements

The author wishes to thank the following individuals for assisting in the development of the manuscript: Laura Kriha for creating the technical forms, and Rita Mosley and Karl Benson for providing a technical review of the manuscript. I would also like to thank the following individuals from Midas International that assisted in the photography and hazards identification: Sam Spitale, Bob Petersen, Neil Casey, Mike Polk, Steve King, Curtis Hazell, and Ray Crane. From the National Safety Council, Neil Cox provided authorization to quote from the NSC's materials.

Chapter 1

Introduction

If an organization is looking for a program to reduce injuries, improve ergonomics, reduce product and property damage, lower their workers' compensation charges and eliminate on the job illnesses, Job Hazard Analysis (JHA) is the answer.

JHA is a special tool for injury prevention that has been used by many organizations that are leaders in safety and health practices. JHA requires an end-product that identifies the basic steps for each job or task, identifies the hazards associated with the job, and develops safe operating procedures to avoid the hazards. When the program is properly designed and utilizes the quality of the finished JHA as its foundation, the organization will be much more successful in reducing injuries and illnesses. Each organization that uses safe working procedures and a job analysis type of program has had success by incorporating it into their safety program.

JHA is a program that helps to educate the supervisor and worker while their job is being analyzed. The hazards associated with the job are identified and steps are taken to remove the hazards. The hazards could have existed for years and not recognized or were responsible for causing injuries. JHA is designed to identify hazards through the team efforts of the supervisors and workers. The first hurdle is for management to recognize that their safety and health program is in need of improving and that they are willing to endorse JHA.

Some refer to job hazard analysis as job safety analysis (JSA), safe job procedures (SJP), and process hazard analysis (PHA). Most of these processes are essentially the same. The titles may vary but the end result is that a safe procedure has been developed for a task or job. For your convenience, we will refer to all of these as "job hazard analysis."

Once the JHA process has begun, it should continue indefinitely. Jobs or tasks are selected and then analyzed by using a step-by-step process to provide a procedure on how to safely perform the job or task. Completing JHAs is not difficult. However, the process requires dedication in order to successfully achieve a safe procedure. There is a learning curve involved that will provide continuous improvement and evaluation, and allows management to measure the time invested versus the results of the statistics attained. Management must be willing to endorse the process. In the early stages of a JHA program, time will be spent performing each analysis. When the analyses are completed, someone must take charge of the project to ensure that the JHA's are correct and accurate.

A JHA program can be used in any industrial setting, construction project, or anywhere else that management has individuals working. Job hazard analysis can be used for something as large as building an airplane to something as simple as making a pot of coffee. Of course, the JHA's required to safely build an airplane would number in the many thousands. The point is that if the principals of the program are followed correctly, there can be benefits to any industry.

Benefits

Job Hazard Analysis is not intended to be a quick-fix to a problem. It is a long-term program that is constantly being updated and modified. Not that a facility will be completing JHAs forever—the JHAs that have been completed will be reviewed and updated where necessary. The forms can also be used for new employee orientations. They can be used for assisting with the investigation of injuries and incidents as well as helping to resolve safety disputes in the plant. There is probably no better means of making the workplace safer than through the use of job hazard analysis.

The specific benefits derived from the JHA process are many. All of the time and energy that is placed in the JHA program will be of value to the organization. The following benefits from JHA can enhance every safety and health program:

- The JHA process documents the safe and exact method of completing a task, which can be used for safety and health training programs.
- Workers are provided with a voice in the safety and health program. The JHA process allows workers to participate in the development of processes, offer their knowledge on how to perform a certain job, and identify the associated hazards.
- Hazards associated with the job are identified during the JHA, so the probability of injury or illness is reduced or eliminated.
- The process of completing a JHA with a worker is regarded as a means to provide the best safety training that can take place. This one-on-one approach re-educates the worker(s) and supervisor as to the correct method in performing the task.
- As safer jobs are being created, the efficiency within the department is improved.
- Supervisors and workers come away from the JHA process more knowledgeable in how to do the job in a safer manner.
- Jobs can be selected from a master list that places them in a priority ranking.
- JHA can influence the purchase of safer equipment and enhance "safety through design" programs.
- Supervisors usually become educated to the means and methods of producing a particular part or completing a task by preparing a JHA with an experienced worker.
- Employee relations are improved as a result of the involvement of the workers and management in the safety and health program.
- Completed JHAs can be used for safety orientations of new or transferred employees.

- Completed JHAs should be reviewed if an injury takes place so that the process can be evaluated and improved upon if necessary. If a hazard was overlooked during the original JHA, steps can be taken to correct the hazard(s) and modify the JHA.

- The use of this program can help in meeting ADA and ISO job description requirements.

- Completed JHA's should be used for safety meetings for the workers and supervisor to review the job at hand and then make corrections or updates if the form requires it. Those jobs that take place infrequently can be reviewed prior to starting them. This refresher will reduce the probability of injury for a job that may not be familiar to the workers.

- Workers' compensation costs should be lowered as a result of fewer injuries occurring in the workplace.

- Hazardous chemicals in use can be scrutinized to ensure that the hazards are corrected in relation to the requirements of OSHA's Hazard Communication Program.

- Ergonomic issues can be brought to light through the completion of a JHA and corrected in accordance with state and federal guidelines.

- Ensuring that safe and correct procedures are being followed enhances workplace quality efforts.

- Housekeeping and workplace image are improved.

- The JHA program satisfies the requirements of OSHA's Voluntary Protection Program guidelines for their application process.

- The results of certain JHAs can help in obtaining funding for job improvements.

- Full use of the program can reduce the number of OSHA violations if the site was inspected.

Regarding accuracy, many safety professionals and plant managers have been required to provide testimony during a deposition or at a trial regarding a workplace injury or product liability lawsuit. Questions asked by attorneys can be loaded with details, and the person being asked has to respond appropriately. If asked if there is a safe way to perform a job, would management have an answer? If asked if management evaluated the job or product for hazards, would management have an answer? If asked if there was an injury associated with a particular machine or process in the past, management would be required to provide a response. In situations such as these, a job hazard analysis can provide a method to prevent the incidents from occurring in the first place. Not that the JHA process is perfect. In fact, anyone can be drawn into a lawsuit. Even in the safest plants, an injury can occur or a consumer could be injured from a manufactured product. However, if the JHA process is used correctly, the probability of injury can be greatly reduced.

Industry should be aware of the fact that as time goes on, the costs associated with injuries, especially medical costs, will continue to increase. It behooves those in industry that choose not to endorse and utilize safety and health programs to consider the impact of injuries on their bottom line. There are some members of management that respond best to economic issues. You can gain their attention and promote safety by talking the language of business: money. Safety has to be looked upon as a means of protecting the profit and losses of an organization. Safety should not be considered a profit-maker but a means of preventing profit loss and saving money for the organization.

Background

It is believed that much of the credit for the discovery and use of job hazard analysis is owed to the steel industry of the early 20th Century. Originally, the program was usually referred to as job safety analysis (JSA). In the 1930's and early 1940's, as other industries were growing, safety programs were being developed in the steel industry. It should be pointed out that there was no OSHA at that time, but some companies realized that the safety of the worker was something that deserved attention, and that their protection was important. Big Steel was growing rapidly at that time, and owners understood the merits of worker safety. Steel was a necessary commodity for the growing economy. Steel was needed for bridges, cars, buildings, airports and for military hardware. With the number of employees in the steel industry now numbering in the hundreds of thousands, workplace injuries and fatalities were increasing. Something had to be done to reduce workplace injuries. Formal safety programs were developed, and job safety analysis was created to focus on the most hazardous jobs. It resulted in great successes. Big steel was responsible for being one of the first industries to require machine guards, utilize lockout/tagout procedures, require the use of personal protective equipment, and provide training for workers.

Big Steel did an excellent job of developing the JHA concept. As a result of JHA, and other safety measures taken at that time, injuries began to decline. It was recognized that safety was good business. Also, the reduction of injuries in basic steel added to the companies' bottom lines. As the JHA program evolved, the JHA process was adopted by other businesses with the same beneficial results.

Commercial Comparisons to JHAs

For those readers that have never heard of JHA, or used JHA in their workplace, the process can be compared to directions provided with most commercial products. The step-by-step description of a process that is the fundamental aspect of JHA is also a part of the instructions provided with many commercial products. For example, a child's swing set, a lawn mower, an electronic game, and frozen dinners all come with a set of instructions. When a parent spends hours putting together a toy and ends up with left over parts, they may have been the victim of poor instructions. Step-by-step instructions should be precise and logically ordered. If gaps are found in the instructions, or the instructions are vague, misleading, or confusing, the quality of the product will be unsatisfactory or the consumer could be injured. Imagine completing a JHA in the workplace and omitting several key work steps. The worker may not simply end up with extra parts—they might suffer a serious injury.

Let us consider one common commercial comparison to JHAs with which most, if not all, of us are familiar--preparing a frozen dinner. To properly cook a frozen dinner, the manufacturer provides preparation instructions for the consumer on the side or back of the packaging. If there are gaps in these instructions, or they are vague or misleading, the consumer could easily ruin his dinner and perhaps injure himself.

The following set of instructions for preparing a microwave dinner are direct, to the point, and easy to understand, as all JHAs should be:

"Since microwave oven wattage varies, cooking times may require adjusting.

1. Remove tray from outer carton.
2. Cut film cover to vent in between chicken and potatoes.
3. Cook on High for 1 1/2 minutes.
4. Using potholders and both hands, carefully remove tray from microwave.
5. Carefully remove film from chicken portion only. Using potholders and both hands, carefully return tray to microwave.
6. Continue cooking on High for 3 1/2 minutes. (If cooking two packages, cook 8-11 minutes.)
7. After cooking, let stand in microwave 1-2 minutes.
8. Carefully remove tray from microwave and carefully remove cover."

The manufacturer of this frozen dinner provides enough detail in the preparation instructions for the consumer to cook the frozen dinner. The instructions also indicate how to avoid injury while preparing the dinner. This same principle is needed in the workplace: Workers must be provided with instructions that describe exactly how to safely complete a task. The completed JHA must provide specific instructions and information on how the job is to be performed, and it must clearly identify the hazards involved.

Other Comparisons to JHAs

The difference between the safe instructions provided by most commercial products and a JHA is that a JHA identifies the hazards involved in each of the steps of the job. Consider the following instructions for changing a flat tire in a driveway:

1. Open the trunk and remove the spare tire, jack, handle, and lug wrench. Place the spare tire near the flat tire but out of the way.
2. Place a block or wedge in front and behind a good tire to prevent movement of the car.
3. Place the scissors jack under the side of the car in the slots provided by the manufacturer.
4. Remove the wheel cover with the lug wrench.
5. Insert the crank handle in the jack and rotate the screw mechanism. Raise the side of the car but do not raise the tire off of the driveway.
6. Loosen and remove the lug nuts and place them in the dish of the wheel cover.
7. Crank the handle on the jack far enough to allow for the removal of the flat tire.
8. Quickly place the good tire on the car and hand-tighten one or more lug nuts.
9. Lower the car slightly and install the other lug nuts. Tighten them down securely with the lug nut wrench.
10. Install the wheel cover and use the jack to lower the car.
11. Remove the scissors jack and place the flat tire, jack, and wrench in the trunk.
12. Remove the wheel block.

What has been provided in the example above for changing the flat tire is not a JHA. What is provided is a procedure for completing the task. Notice that the hazards involved in each of the steps of the job was not identified. The correct procedure, which would include the hazards and how to avoid them, was also omitted. To assist readers in the use of the correct JHA forms and process, a completed JHA on changing a flat has been included in Figure 1.1.

The JHA Program

Defining the Job or Task

Within the JHA program, there are words and phrases that are unique to the process. A key word is "job," as used in job hazard analysis. The word "job" is usually associated with an occupation such as an electrician, auto mechanic, welder, carpenter, punch press operator, carpet installer, etc. In each of these job classifications, or occupations, there are tasks or projects that the worker performs. Each task has a series of steps from beginning to end. The objective of the job hazard analysis is to observe the worker and to record the process of one individual task from beginning to end.

The term "job," in the context of the job hazard analysis program, represents a sequence of definite steps or separate activities that together accomplish a work goal. *It does not represent the occupation of the worker.* As an example, a forklift operator does not just operate a forklift truck; he has various duties to perform. These duties, or jobs, go beyond just picking up a pallet and moving it. Some of the tasks or jobs that a forklift operator could be involved with could include: replacing a propane cylinder, charging a battery, stacking pallets, hand loading cartons on a pallet, unloading a trailer and so on. Each of these tasks or jobs is to be analyzed for hazards. The occupational or work classification of a forklift driver does not allow for all of these jobs to be done at one time. The analyses of all of the tasks involved in the operator's occupation could take months to complete. Each step of a specific job may have one or more hazards that can cause injuries or illnesses to the worker. Some of the steps within the job may have serious or life-threatening hazards. Other steps may contain no hazard at all. The reader may be asking the question" what can be done by the worker or management to make the job safer?" In the following sections, we will address this question.

Management's Role

It is essential that management identify the hazards, or the means by which injuries could occur. If the hazards are unknown, nothing can be done to correct them. It is necessary that management identifies and corrects hazards in the workplace so that the hazards are eliminated. Job hazard analysis satisfies this requirement. When each step of the JHA is observed and analyzed, the sources of injury can be identified.

When observing the job steps, there may be a need to improve the safeguards in the working area. This could include installing ventilation, installing or replacing a guard, providing personal protective equipment (PPE), and possibly repositioning or relocating a machine to provide more working space. Guardrails may be needed to protect the worker

JHA Format Which Should Be Used To
Complete Each Job Evaluation
"Changing a Flat Tire in a Driveway"

Basic Job Steps	Hazards Present in Each Job Step	Correct and Safe Procedures For Completing the Job
1. Open trunk, remove good tire / tools.	1. O – overexertion while working and lifting in trunk.	1. Open the car trunk and remove the good tire, jack, jack handle, and lug nut wrench. Place near the flat tire for easy access but not in a position to obstruct removal of the flat tire. Bending over and lifting from a trunk can result in a back injury. Do not pull or jerk the new tire while lifting it.
2. Block the tire.	2. No Hazard	2. Place a block or wedge against the front and rear of a full tire to prevent movement of the vehicle.
3. Properly place jack and remove wheel cover.	3. CB – fingers can be caught between the wheel and lug nut wrench.	3. Set scissors jack in place under the frame of the car in the spot designated by the manufacturer. Consult the owner's manual if necessary. Keep the hands and fingers clear of the pinch point between the tire and lug nut wrench when prying off the wheel cover.
4. Raise vehicle.	4. O –overexertion while turning handle.	4. Insert the crank handle into the jack and slightly raise the vehicle. Check to ensure the jack has been properly placed. Turn the handle (usually in a clockwise direction) and lift the vehicle but do not lift the tire off of the driveway.
5. Remove lug nuts.	5. O - there could be a strain or overexertion while removing lug nuts.	5. Place lug nut wrench securely on each lug nut. Press down in a counter-clockwise direction and remove the lug nuts. Place lug nuts in wheel cover.
6. Raise vehicle.	6. O – overexertion while turning handle, and lifting tires SB – falling vehicle.	6. Raise the vehicle by turning the handle far enough to lift the flat tire off of the pavement. Be alert because the vehicle could possibly fall at this time. Remove the flat tire and place out of the way. Immediately place the good tire on the vehicle. Add the lug nuts by hand tightening as soon as possible. Avoid injury to the back by safely lifting the tires. Use the arm and leg muscles and do not rotate / twist the back while lifting.
7. Slightly lower car; tighten lug nuts.	7. CB – fingers caught in wheel cover.	7. Lower the car slightly so the tire is touching the pavement. Use the lug nut wrench to tighten the nuts. Do not under-tighten or leave any loose nuts; the tire could fall off of the car while being driven. Install the wheel cover correctly and securely and keep the fingers clear of the pinch point between the wheel cover and tire.
8. Lower vehicle; replace tools / tire.	8. O – the back could be injured while lifting.	8. Use the crank handle and lower the car all the way down. Remove the handle and jack. Return the flat tire and tools to the trunk of the car. Remove the blocks from the wheels. Use the legs and arms while lifting to avoid a back injury. (Don't forget to have the flat tire repaired.)

Hazard Selection for the middle column: SB = Struck By; CW = Contact With; CBy = Contacted By; CB = Caught Between; SA = Struck Against; CI = Caught In; CO = Caught On; O = Overexertion or Repetitive Motion; FS = Fall At the Same Level; FB = Fall to Below; E = Exposure to Chemicals, Noise, etc.

Figure 1.1 JHA on Changing a Tire

from the lift truck traffic. Perhaps a hoist is needed to prevent a worker from lifting or manually handling heavy, awkward objects. These are items that are usually identified or "uncovered" during a JHA. And these are some of the reason for incorporating JHA into a safety program: Without listening, observing and asking the worker any questions, how would management know that these form of improvements are needed? Unfortunately, injuries or illnesses to workers at specific machines or operations are usually the trigger for management to make improvements to the workplace. Job hazard analysis is intended to recognize workplace hazards and correct them before an injury or illness takes place. Incorporating JHA in a safety program can be one of the best safety and health program decisions an organization can make.

Another means of providing safeguards for the worker during a JHA is to recommend a change of work procedure. Perhaps the worker is not bending at the knees and keeping his back straight when lifting a load. Perhaps he is overloading carts that require constant pushing or pulling into place. Perhaps the worker is welding and is not placing the flexible exhaust duct near the section being welded. The JHA is intended to identify these dangerous and sometimes inefficient methods and replace them with the proper procedures. The supervisor has the duty to look at the conditions with which the worker is faced, and to decide whether or not there are correct work methods and potential environmental issues to consider. If these conditions, procedures, and exposures are not discovered and corrected, injuries and illnesses are sure to follow.

Personnel Involved

Most organizations take time to consider who will handle the task of observing, discussing, and writing the correct information on the JHA forms. The usual candidates for this task are the first-line supervisors. Most safety programs place the first-line supervisor as the key person to carry out the various components of the safety program elements. The employee's immediate supervisor is usually the most desirable choice because:

- Usually the supervisor and the employee performing the task know each other well and see each other every working day.

- The supervisor usually knows the person well and has knowledge of his hobbies, family members, etc.

- The supervisor usually has a good idea of how the product is produced and how the machines are operated.

- The immediate supervisor knows who would be the most helpful to complete a JHA when selecting someone to assist.

- The immediate supervisor is familiar with the past injuries in the department.

- The immediate supervisor is usually the one that will be on a bonus or safety objectives program.

- The employee usually takes direction from his/her immediate supervisor. If there is a question of changing a work rule or procedure, the supervisor is the best person to initiate this action.

For the most part, the use of supervisors for this process has proven successful over the years. However, other individuals may be assigned the task of completing JHAs, depending on the needs or resources of the company. In some organizations, the quality control person is assigned the duty. In others, the one-on-one method is not used and a group of individuals is assigned the task to periodically complete JHAs through consent. The list of individuals used in the JHA process is long and there are many combinations to select from, but individuals frequently included are:

- Personnel managers
- Human resource managers
- Dedicated safety team leaders
- Safety managers
- Safety committee representatives
- Summer interns
- Quality assurance managers
- Employee focus groups
- Machine operators
- Maintenance personnel
- Safety steering committees

It is up to management to determine who will be the person or persons to complete the analysis. This point will be repeated again in this book, but one person should not complete any JHA without including the employee. Just assigning one person to this task will not be of benefit to the safety program. Without the input of the worker, the JHA program will be a hollow one.

Required Paperwork

The JHA process generally requires three types of forms to complete:

1. To get the JHA process started, a job list is created. The job list (Form #1) is necessary to list and identify by number all of the jobs in the facility. The jobs can be listed by occupation or by priority. Without a job list, the program cannot move forward. The job list will be discussed in detail in Chapter 2.

2. "Rough drafts" (Form #2) of the final paperwork are completed by the supervisor while at the work area with the employee.

3. When the supervisor has completed the JHA, he will submit the final copy (Form #3) to someone for approval. If the form needs work, the supervisor will have to make corrections. Once the form has been approved, it will be placed in a binder, online, or posted near a workstation.

JHA forms #2 and #3 are divided into three columns, and the following information is needed in each of the columns:

- Basic Steps: In the "Basic Steps" column, provide a very brief description (less than six words, if possible) of the step being performed. Do not enter information regarding the how, where, when or why regarding the basic step.

- Key Hazards: In the middle column of the form, there are 11 key hazards that are identified with the following key indicators: CB for Caught Between; CI for Caught In; SB for Struck By; CO for Caught On; CW for Contact With; CBy for Contacted By; FS for Fall-Same Level; FB for Fall to Below; O for overexertion; SA for Struck Against; and E for exposure. The hazard identification symbols should be located at the bottom of the forms for ease of use. Once supervisors, or those that use the forms on a regular basis, memorize the abbreviated hazard keys, they will complete these entries from memory. Another benefit from using the key indicators is that the identification of hazards while walking through the plant or while completing JHA becomes easier. The awareness level for identifying hazards will definitely increase.

- Analysis: In the third column, which is on the right side of the forms, include information on the basic step, the potential for injury or illness which is identified in the middle column, and the description of how to complete the basic step correctly without being injured. The correct amount of information that is entered in this column will depend on the writing skills of the person completing the form.

How to complete the written JHA Forms #2 and #3 is discussed in detail in Chapter 3.

The JHA Process

The first step in the JHA process is to select a job for analysis. A job list must be prepared to identify all jobs at the work site. From this list, jobs are selected based on a priority basis. Of all the jobs listed, the ones that have a history of causing the most serious injuries or a fatality should be chosen first. It is strongly recommended that all jobs, regardless of priority identification, be placed on the job list.

The next step is to observe each job step being completed and take specific notes of the process. Each step has to be analyzed so that any source that could cause an injury or illness can be identified. Each step must be in the order of how the job is performed and the hazards that are a part of the job must be identified.

In the process of identifying the basic job steps, the task is evaluated for the hazards involved. Can the worker: be struck by an object, strike against an object, make contact with a chemical, be caught between two objects or be contacted by a hot part or object? Can the worker have a foot caught in a hole, experience overexertion, fall at the same level, fall to a lower level, be caught on a moving part or be exposed to a toxic chemical?

The next step is to combine the basic job step along with the hazards that are present in the job step and develop a procedure containing the what, how, where, and why of the job. It should be noted that the basic job step does not identify the where, when, why, how, or who. The basic step only identifies "the what." This section of the completed JHA is deliberately kept brief. In the completion of the JHA, the section that combines the basic step and the job hazards will require more detail. This part of the JHA provides specific details on exactly how to safely perform the job.

Each step of the JHA process will be discussed in detail in following chapters, but the process can be summarized as follows:

1. Select a job from the master list; assign job's priority.

2. Record the basic job step and be brief in the identification process.

3. Record the hazards that the worker would be exposed to in each job step.

4. Complete the JHA by combining the basic step and the potential hazards associated with the job.

JHA Methods

The question that usually arises is what method or methods should be used to complete a JHA? Four different systems exist that allow for an analysis of a job.

1. The one -on- one observation method.

2. The group discussion method.

3. The recall and check method.

4. The absentee method.

A discussion of the pros and cons of each method follows, but it cannot be stressed enough that the quality of the completed documents is very important in the JHA process. Take the time to do it right. It is better to produce fewer JHAs that are accurate and correct than to produce many documents that are inaccurate and incomplete.

One-on-One Observation Method

The one-on-one observation method is generally the preferred method of completing a JHA. The JHA program encourages management to work with the employee to develop safety guidelines and to identify hazardous conditions. This method requires that the supervisor choose an experienced worker that would be able assist in providing the descriptive step-by-step information needed for the program. In this method, the supervisor selects the job he/she wishes to analyze and then chooses the worker he will collaborate with on the JHA.

The National Safety Council, in their *Accident Prevention Manual*, states "to do a job breakdown, select the right worker to observe—an experienced, capable, and cooperative person who is willing to share ideas. If the employee has never helped out on a job safety analysis, explain the purpose—to make a job safer by identifying and eliminating or controlling hazards—and show him or her a completed JSA. Reassure the employee that he or she was selected because of experience and capability."

The key is to observe the worker or workers while they are performing the job or task. It is also important to listen to the worker as they convey the correct procedure verbally while completing the task. When choosing a particular JHA for review, include the worker(s) in the selection process when possible. It is appropriate to review a job list with a worker and let them know what job has been selected for review. The employee may be able to recommend a better job to evaluate based on their experience. Do not overlook the value of the input that can be provided by a worker. Also, it is recommended that the name of the worker(s) be added to the completed JHA. It is advisable that the employee be asked about adding their name to the document. The majority of workers will not mind helping to develop safer working procedures.

Pros

The one-on-one method promotes learning on the part of the employee(s) and the supervisor. There are supervisors that have worked in a particular department for years and do not know how each machine operates. When the operator of a machine or the worker that is completing a certain task explains what they are doing, learning takes place. Much can be learned from observation, but more is learned from observing, listening to the comments from the worker, and asking questions about the process. Supervisors definitely learn new things about the job after correctly completing a job hazard analysis. Usually the supervisor comes away from a completed JHA with a better understanding of the dangers or hazards associated with the task itself. There is sometimes a better appreciation of what the worker experiences while completing the task. Perhaps the job involves more lifting, bending, twisting, awkward posturing, or even less job efficiency than first realized. All of this comes about by taking the time to listen, observe, and ask the appropriate questions.

This method also benefits the worker in several ways. First, the worker is being asked to assist in developing a safe procedure. The new or improved procedure will benefit him as well as other workers on the same machine or process on other shifts or in other facilities. Next, the worker is reinforcing his knowledge of how to do the job correctly. There may even be a sense of pride for the worker when he has a chance to demonstrate his knowledge of the job. He may have been doing the same job incorrectly for years and along comes an interested supervisor to help improve working conditions. Many work improvements related to machines and processes are appreciated by workers. Ergonomic improvements are particularly welcome. Any time that the working conditions at a workstation can be improved, the worker will be grateful for the corrections. Being that most of the complaints from employees during safety meetings identify problems with working conditions, management must demonstrate their concern by altering these conditions for the better.

Cons

A potential weakpoint in the one-on-one observation method is that jobs that are performed infrequently can easily be overlooked. They can be left off of the master list, and when they are performed, the supervisor may not have the opportunity to conduct an analysis. An example of this infrequent performance of a task was at a metal stamping plant that rolled 250 pieces of a special exhaust part every 6-12 months. The time-span between the use of the machine that rolled the metal sheets was dependant on the amount of units sold. It was very difficult to plan a JHA on this particular task. It took a while for a supervisor to schedule a JHA on this job, but fortunately he was assisted by the plant manager regarding an estimated production date.

Another difficulty in using this method is for the jobs that are performed off site. Employees that are working in remote areas do not usually have all of the company resources available to them. It is possible for a member of management to be available in remote field operations if necessary. However, the one-on-one method of observation may be difficult to achieve.

The Group Discussion Method

The group discussion method is the next best method of conducting a JHA. This method involves the assembly of a group of supervisors and workers to discuss a particular job and arrive with a completed JHA. It is advisable that everyone in the group be familiar with the job because they can add knowledge to the process. The group will interact and draw from their experience to arrive at the completed JHA.

The following points should be recognized when using the group discussion method.

- One person should act as the discussion leader. The discussion leader should be skilled in this process and must keep the group focused on the job at hand. The leader may wish to designate a scribe to record the decisions of the group. The leader may wish to use a chalkboard, flip chart, or overhead projector to record the decisions of the group. The group can focus on the solutions and not be concerned with taking notes.

- Individual ideas from within the group can be tested against the experience of others. If one person overlooks a key point, someone else can offer comments to make it accurate. For the most part, there is usually agreement among these groups, and the finished JHA will receive a significant amount of input.

- This method is suitable for the infrequently performed jobs because the attendees will most likely know the details associated with the job. Those jobs that take place at a remote site are also good candidates for this method. This is another situation where the worker in attendance with the right experience can be of enormous assistance.

- The plant manager that utilizes good time management skills can schedule these meetings during periods when the plant is shut down. There may be times when the plant has a scheduled shutdown for several different reasons. This idle time can be used to gather together select employees and supervisors to discuss and complete specific JHA's.

- Safety meetings can also be a good time to schedule group discussions for JHA's. In any of these group meetings, the discussion leader should control the number of individuals in the room. A large group attempting to reach consensus can be difficult. Valuable time can be lost while a large group of people attempt to agree on key points. If the group is being coordinated, the JHA can be completed within the framework of the meeting time.

Pros

As with the one-on-one observation method, the group discussion method provides training and reinforcement for those in attendance. The sharing of experience among the group helps to train everyone in the room. In fact, there may be more knowledge gained by some individuals than they expected to receive. This gaining of knowledge also promotes acceptance among the group. The group discussion method increases the learning of the job procedures. There may be more knowledge gained through this method than the one-on-one observation method because more ideas are discussed.

One of the strongest benefits of this method is that the completed JHA does not have to be reviewed or evaluated by anyone else. Once the group completes the process, the JHA should be ready for formal processing.

Cons

It may be difficult to assemble a group of people at a machine or operation to discuss the safe procedures. Experience has shown that gathering at a machine can easily become unwieldy and disruptive. The recommended means of using this method is for the group to meet in a room to discuss the JHA. Also, control of the group may take more effort than writing the analysis.

The Recall and Check Method

The recall and check method requires the supervisor to think about a certain job or operation and develop a JHA based on the supervisor's ability to recall the job processes and then check them for verification. The supervisor must use his memory and recall to arrive at the correct conclusions. He would then show the JHA to another supervisor or worker for verification. At this point in the checking process, the progress of the JHA is somewhere near the completion process, but not as complete as a JHA completed in the group discussion method.

Pros

This method has an advantage because it can be used to analyze the infrequently completed tasks as well as the jobs that take place in remote areas. However, the supervisor would have to be very experienced in any of the jobs to be able to accurately recall all of the details associated with a job. If the organization has the time and resources, and can assign more than one supervisor to the writing of a JHA, the quality of the completed project would be much better when using this method.

Cons

Employees may not have a say in the entire development process. The supervisor that completes a JHA with this method may not use an experienced worker to assist in the "check" portion of the process. Keep in mind that employees are supposed to be on the receiving end of JHAs. As a result, they should have input into the process. From the standpoint of the development of a JHA, employees are less likely to violate safety practices that they had a hand in developing.

The Absentee Method

The absentee method is appropriately named for the creativity of a supervisor that is required to complete one or more JHAs and proceeds to take them to his home for completion. Management should discourage the use of this method unless there is no other way to complete JHAs by using the other three methods. The means of salvaging the work put into these drafts of JHA's through this method would be to allow others to review and offer corrections where necessary.

Cons

There is no learning in this process. The employee is not a part of the development of the analyses. In many cases, the supervisor has the knowledge of the job but fails to include the input from anyone else.

Summary

Unlike other elements in a safety program, JHA takes a little extra time to discuss and outline. It is not a program that is usually learned overnight. Readers are encouraged to start developing a mental picture of how they will incorporate this process into their over-all safety and health program. Some may comment that "there is no more room in our program. We have enough to do just getting the production out the door than to be concerned with a new program that will take awhile to learn and execute before we see progress". The comments are valid. Industry is being downsized, right sized, and as some say, cap-sized. Many organizations have cut back staff to where some supervisors are doing the work of two or more people these days. Staff is stretched very thin.

However, this trend has not helped in making the workplace safer. Staff cuts may have made the workplace less safe. It should be pointed out that some organizations have been using the same programs for the past x years and have not even made a real dent in their injury totals. The trend sees workers' compensation rates rising, mostly as a result of increases in medical costs. In addition, OSHA is still a force to contend with for those organizations that do not place safety in a high priority. The current push nationally for greater ergonomic controls will not disappear.

Through the use of job hazard analysis, organizations can achieve more than they know. The benefits derived from this program are many. Subsequent chapters in this book will provide readers with enough information to provide many improvements to their work-places. Yes, it does take a while to learn how to properly complete a JHA. Yes, management is being stretched even thinner these days. Yes, if production does not go out the door, no one may have a job. All of these statements are true, but the JHA program can help all of the concerns. Through the correct use of this program, losses due to injuries, damaged product, and damage to the building will be reduced. Housekeeping, better employee relations, and improved work procedures are another benefit of correctly using the program. In fact, there is probably no better means of improving safety and health performance than through the use of JHA.

Chapter 2

The Job List

In the JHA process, it is necessary to develop a job list, which may also be referred to as the master list (Form #1). This list identifies all of the jobs that are to be analyzed. Management should consider all jobs as having some danger or hazards in them. Every job in a facility or at a job site should be added to the master list of jobs. The job list is a means of outlining all of the jobs in a facility so that everyone knows what jobs have to be analyzed.

Although this list may be duplicated as necessary, the master list must be a controlled document. The list must be monitored because it is a "living" document, in that jobs can be added, deleted, and checked off as they are completed. Once the initial training has taken place, a listing of jobs for each machine, operation, department, worksite, and plant should be developed before the JHA program proceeds.

Without the benefits of having such a document, there would be chaos in attempting to coordinate job selections for completing a JHA. Imagine being asked to make a trip by car to a town in a far away state; a place that you have never been to. Also imagine starting out without having a road map to provide instructions on how to get there. This is what the JHA job list is all about—it is the road map to get you where you are going.

The selection of jobs from this extensive listing allows supervisors and employees to select the jobs management has assigned to them for analysis. In many cases, jobs are assigned and the process of job selection may be directed to the choice of one job from several that provide the same degree of risk. Figure 2.1 illustrates a blank job list that may be used as a template for developing your own.

Benefits of the Job List

The creation of a comprehensive job list is the stepping stone to the JHA process. The job list is important for several reasons:

1. It puts the JHA program into motion by giving it direction. Without a complete listing of jobs, the process would be hard to organize. Jobs are chosen and assigned from a list. If supervisors chose their own schedule of JHA's, there could easily be several individuals selecting the same job. Not that the selection of the same job should be discouraged, however. There can be improved learning when similar JHA's

Job Hazard Analysis Master Listing Of Jobs
Form #1

JHA Number:	Title Of Job Hazard Analysis:	Job Assigned To:	Date Assigned:	Completion Date:	Review Date:

Figure 2.1 JHA Form #1

are completed then compared. But because time is so important to managing a safety program, it is more expedient to assign jobs to an authorized individual, who then properly completes the analysis and shares the evaluation with others.

2. The job list allows employees and supervisors the opportunity to compare notes and discuss the variety of jobs that exist in a department. The combination of management and labor working together will result in a safer department.

3. A listing of jobs allows for specific selections of jobs. Each supervisor brings a certain level of skill and knowledge to the job. Someone who is new to the department can't possibly know how machines and processes work. To analyze a job, one must have a basic understanding of what is going on in that job. Some individuals have a real grasp of how a certain machine operates. Others may have a difficult time completing the same evaluation because they lack the job knowledge. When jobs are being chosen from a list, management can be served better by allowing individuals the opportunity to choose one of several jobs based on their job knowledge and skills.

Creating the Job List

To create and prioritize a job list, it is necessary to evaluate past losses and near misses. A large part of this work may already have been accomplished through your existing safety and health program. This process of information gathering is a definite asset not only to the development of a job list, but to the effectiveness of your current safety and health program. Reviewing relevant documentation and consulting workers is the best way to evaluate the current condition of the workplace.

Specifically, a search should be made of company documents, such as workers' compensation loss runs, OSHA #200 logs, insurance injury investigations, and department incident reports. Also, and perhaps more importantly, workers on the various machines and operations should be consulted because they can contribute significantly to the information needed to develop a job list. The workers know the history and background of each job—after all, they live with these jobs every day. As a means of gathering information, ask the workers to identify all of the injuries, illnesses and incidents that have taken place at their jobs. From this information, the jobs that produced these injuries will be placed on the job list in the priority they deserve. It would be unfortunate if management developed a list without consulting the very people that run the machines and produce the product. It is unlikely that management knows of every close-call and non-injury incident that takes place.

The resulting job list should be as complete as possible. A thorough search will help identify those jobs whose hazards may seem insignificant, but have significant associated risks nonetheless. Although it may seem odd to include a job as simple as sweeping the floor on this job list, consider the following example, in which both jobs were considered routine, simple, and not hazardous:

In the first incident, an experienced technician was sweeping the shop floor with a large push broom. He had one hand on the upper part of the broom handle and one hand several feet lower on the handle. In the process of sweeping he was backing up to sweep over a section of the floor that was previously swept. He did not realize that a two-inch diameter section of exhaust pipe was sticking out of a vice. The worker cutting the piece of pipe with a hack saw earlier in the day failed to remove it from the vice. While backing up with a rearward motion of the broom in a forceful motion, the small finger of his left hand made contact with the razor-sharp edge of the pipe and severed the first digit.

The injury could have been prevented by first removing the pipe from the vice. Parts should never be left in a vice, especially pieces with razor-sharp edges. It is not unusual for someone to walk into a part or object left in a vice and experience an injury. Also, one should always look in the direction that they are moving and have a general awareness of the area in which they are working.

In the second incident, a mechanic was using a squeegee to move water off of a shop floor after the floor had been hosed down for cleaning. It was at the end of the day, and the established procedure was to raise all of the in-ground vehicle hoists to effectively do a good job of sweeping and using the squeegee. All of the hoists were raised as far as they could be lifted. While busily involved in the process of moving the water into the driveway with the squeegee, the technician failed to pay attention to the hoists, which were at head level. He walked into a hoist and severely lacerated his forehead. The injury required a dozen stitches to close the wound. He stated that he forgot that the hoists were in the air and at head level.

The two examples of injuries cited above represent basic jobs that are a part of many operations. Many organizations would not consider adding these jobs to its job list. However, the hazards associated with each job demonstrate the need to include all jobs on a job list, regardless of how dangerous they are or how safe they may appear. Never underestimate the need for performing a JHA on even the most simple and basic task.

However, jobs that result in more frequent injury and illness deserve to be evaluated on a priority basis to prevent occurrence. Management can select the most dangerous jobs first and then proceed to include other jobs. A JHA program can only be effective when jobs are analyzed on a priority basis. Once the job list has been developed, priority should be assigned to each job according to the following order:

- Jobs that have the potential for producing serious injury or a fatality;
- Jobs that have consistently produced injuries and illnesses;
- Jobs that have resulted in cases of high severity in injuries;
- Jobs that have the potential for causing a high frequency of injury or illness;
- Jobs that are new, or involve new machines or processes;
- Jobs that have a high incidence of close calls, property and product damage; and
- All other jobs.

The job list should be kept as up-to-date as possible. When injuries occur in a specific job that is not currently on the list, the job should be added and the job list reprioritized. Be sure to add and remove jobs from the job list as machinery is introduced or removed from the workplace. It is recommended to include jobs on the job list as they are created, and to complete a JHA on new jobs before they can become a potential hazard to workers.

Identifying Jobs for the List

The placement and identification of the jobs on the job list is very important. Each task that an employee performs as part of his job duties must be named and assigned a number or letter scheme so that when the JHA is performed, the task can be quickly and easily identified.

This is best demonstrated through an example. Let us use the job classification of a hypothetical warehouseman. In the scope of his duties, he will most likely perform the following jobs:

- Load and unload trailers with the use of a forklift;
- Pack and unpack cartons of product;
- Assemble and disassemble product racking;
- Manually load and unload trailers;
- Use a power drill, circular saw, hammer, screw driver, and strapping (banding) machine;
- Operate a forklift for a variety of purposes;
- Operate additional powered industrial trucks such as a pallet jack and an order picker;
- Sweep the floor; manually or with a powered sweeper;

He may have to perform the following:
- Use a fire extinguisher;
- Use a chemical spill clean up kit;
- Repair broken pallets.

To prepare a job list, each of the jobs listed above must be assigned a name and number or letter identification. Let us assume that this hypothetical warehouse in Pittsburgh is one of seven in this growing corporation, and has decided to use the JHA program to improve their safety performance. As a result of the above information, the JHA list may identify the facility and jobs in this way:

Pittsburgh Warehouse Job Safety Analysis List

Job Classification: Warehouseman

PW 1: Loading a trailer with a forklift

PW 2: Unloading a trailer with a forklift

PW 3: Packing cartons for shipping

PW 4: Unpacking cartons received at the dock

PW 5: Assembly of product racking

PW 6: Tear-down of fixed racking

PW 7: Manually loading trailers from extended conveyors

PW 8: Manually unloading trailers onto extended conveyors

PW 9: Safe use of powered and manual hand tools

PW 10: Safe operation of a forklift

PW 11: Safe operation of an electric pallet truck

PW 12: Safe operation of an order picker

PW 13: Manually sweeping the warehouse

PW 14: Using the power sweeper

PW 15: How to properly use a fire extinguisher

PW 16: How to use the spill cleanup kit

PW 17: How to repair broken pallets

PW 18, 19, 20, etc. etc.

(PW=Pittsburgh Warehouse)

This example demonstrates the method that can be used in the development of the job list. Without the job list, there will be confusion in the JHA program. In this example, a warehouseman has the potential for injury in a variety of job tasks. The list could be longer or shorter depending on the work and assignments at the warehouse.

If this were an actual job list, it would be important to select jobs to be analyzed from the list based on a hazard priority system. A selection would be made from the list by choosing the most dangerous jobs first, which depends on the injury and loss history in the Pittsburgh warehouse.

All of this could be a part of a larger scheme, which would involve other employees in other job classifications and with other JHA titles. When considering the various occupations or job classifications at the Pittsburgh warehouse, there could easily be a master list of 100 or more jobs. The job list would vary depending on the equipment and tasks in the warehouse. It is important to develop this listing of job assignments and the tasks for which each employee would be responsible. Each department should be responsible for developing their own lists, which are then collected into one master list.

An organization has the choice of including jobs on the job list in any manner they wish. For example, jobs can be entered on the list by priority, listed by occupation or job classification, or listed randomly (the "shotgun effect"). Management may arrange the list according to company preference, as long as the jobs chosen for analysis can be easily identified by the JHA documentation and those performing the analysis. The jobs do not necessarily have to be listed in order of priority, but when it is time to perform a JHA, those jobs with the most associated risks should be analyzed first. However, there are exceptions to every rule. For those supervisors that have never conducted a JHA, management may wish to choose to analyze a job that is not too difficult first so the learning curve of the individual can be improved. This particular job may or may not be as hazardous as other jobs on the list—but it could help in learning the process.

Sample Job Lists

The following extensive lists are meant to provide readers with enough details to assist them in the development of their own job lists. The three occupations or job classifications used for the following lists were selected because there are forklift operators, warehouse persons, and a maintenance staff in nearly every warehouse or distribution center. Note the listing by classification of all the jobs that are a part of the workers' duties.

Warehouse Listing of Jobs Required for a JHA

Forklift Operator:

1. Loading trailers
2. Unloading trailers
3. Hand-stacking pallets
4. Repairing pallets
5. Charging batteries
6. Cleaning batteries
7. Completing a check and inspection of the forklift
8. Spotting loads in upper tiers of the racking
9. Applying shrink-wrap to pallets of product
10. Placing a new roll of shrink-wrap on the shrink-wrap machine
11. Lifting loads outside off of a flatbed truck
12. Correct use of the eye-wash station
13. Correct use of the full-body shower
14. How to safely clean up a chemical spill
15. Use of automatic openers and spotting a load on the mezzanine
16. Correct use of a fire extinguisher
17. Using the plastic banding machine on pallet loads
18. Changing propane tanks on the forklift
19. Correct use of a safety harness

Warehouse person:

1. Sweeping the warehouse floor
2. Operating the powered floor sweeper
3. Manually handling small parts
4. Storing product on the mezzanine
5. Disassembling metal racking
6. Assembly of metal racking
7. Use of a portable power saw to cut wooden sheets

8. Use of the compactor
9. Use of the banding machine on pallet loads
10. How to use a portable fire extinguisher
11. Sweeping out a trailer
12. Patching a hole in a trailer floor with wood sheets
13. Packing UPS orders for shipment
14. Operating the conveyor
15. Manually unloading product from a truck or trailer onto a portable conveyor
16. How to use a spill cleanup kit
17. Cutting metal banding off of shipments
18. How to safely drive a stock chaser
19. How to inspect a stock chaser
20. How to safely operate an electric pallet truck
21. How to inspect an electric pallet truck
22. How to use an eye wash station
23. How to install portable jacks under the nose of a trailer
24. Using the paper/corrugated cardboard baling machine
25. Correct use of a NIOSH approved dust mask
26. Safe handling of shipments of chemicals at the small parts department
27. Safe use of the staple gun
28. Manual handling of totes
29. Safe use of a rolling ladder
30. Using the material elevator

Maintenance worker:

1. How to safely shovel snow
2. How to mow the lawn(s)
3. Repairing a broken rolling ladder
4. Changing the ballast in a light fixture
5. Adding a guard to a machine
6. How to safely use lockout/tagout procedures
7. Setting up the a-frame for confined space entry
8. How to safely test the air/environment in a confined space
9. Safe use of weed killer on the lawn
10. Safe use of the scissors lift
11. How to inflate pneumatic tires
12. Replacing a broken window

13. Safely wiring an electrical fixture
14. Installing carpeting
15. Repairing a dock plate
16. Installing decking on a mezzanine
17. Patching a leak in the roof
18. Repairing an overhead door
19. Installing sections of portable conveyors
20. Replacing a motor on a conveyor
21. Replacing a switch in a circuit breaker box
22. Repairing the shrink-wrap machine
23. Installing overhead lights
24. Safe use of electric power tools
25. Painting yellow lines inside and outside of the building
26. Correct use of an organic vapor respirator
27. Replacing a tire on a forklift
28. Safe use of an extension ladder
29. Installing a plumbed eye wash station
30. Installing a safety shower
31. Replacing a standard door
32. Installing an emergency alarm on an exit door
33. Installing protective barriers around equipment to prevent lift truck damage
34. Sprinkler testing
35. Installing a new sprinkler head
36. Safe use of a bench grinder
37. Safe use of a hedge trimmer
38. Repairing damaged hand railing
39. Safe use of a fire extinguisher
40. Installation of racking corner protectors

The previous job listings have focused on warehousing. The following job lists focus on construction, a power press at a metal stamping plant, and an automotive repair shop. These lists are intended to serve as reminders of how jobs are to be identified and then added to the list.

Jobs at a Typical Construction Project

1. Correct method for wearing a safety harness
2. Erecting a scaffold

3. Installing a wooden float for high work
4. Using a shovel to dig a trench
5. Using a tag line to control suspended beams
6. Safe use of a powered nail gun
7. Erecting 2 x 4 hand railing
8. Erecting wire rope as hand railing
9. Safe use of a wheel barrow
10. Safe use of a portable power saw
11. Safe use of a hand saw
12. Safe use of a hack saw
13. Applying 4 x 8 sheets to roof trusses
14. Correct installation of roof trusses
15. Unloading a pickup truck of cement bags
16. Safe use of an air impact gun
17. Correct use of a cutting torch
18. Removal and installation of compressed gas cylinders
19. Correct use of a fire extinguisher
20. Correct use of an extension ladder
21. Removing a scaffold
22. Installing windows
23. Safe removal of asbestos from a boiler
24. Safe removal of asbestos from a steam pipe
25. Applying tar to a roof with a mop
26. Connecting steel beams with bolts
27. Sorting steel beams using a mobile crane
28. Performing a daily inspection on a mobile crane
29. Laying concrete blocks
30. Mixing mortar
31. Correct use of a NIOSH approved dust mask
32. Pouring concrete into forms
33. Laying wood footers for forms
34. Correct method of building a job ladder
35. Laying felt paper on a roof
36. Unloading a truck of felt paper
37. Safe use of a winch
38. Correct procedure for cleanup of a chemical spill

39. Safe use of a table saw
40. Changing a blade on a table saw
41. Correct use of an eye wash station
42. Hanging a door
43. Safe use of a backhoe
44. Safe welding procedures for small jobs
45. Sweeping a recently poured concrete floor
46. Installation of dry wall
47. Application of plaster to walls and ceilings
48. Laying carpeting
49. Carrying mortar
50. Applying solder to pipe joints

Jobs Related to a Power Press Operator

1. Inspecting the machine before start up
2. Safe use of the die changing cart
3. Using a hoist to place a coil of steel
4. Feeding a coil strip into the machine and set up
5. Operating the machine
6. Safe method of adjusting the guards
7. Removal of scrap sheets
8. Using the powered pallet truck to move full hoppers from the machine
9. Completing a daily inspection on the overhead hoist
10. Steel coil and reel set up
11. Removing dies
12. Installing dies
13. Floor sweeping and clean up

Jobs Related to an Automotive Repair Shop

1. Safe use of fore and aft hoists
2. Changing oil on a car
3. Safe operation of a brake lathe
4. Safe removal of tires
5. Shoveling snow
6. Safe use of a drum/rotor lathe
7. Installing an exhaust system
8. Removing an exhaust system

9. Removal of shock absorbers
10. Installation of shock absorbers
11. Balancing a tire
12. Installing a tire on a rim
13. Removing a tire from a rim
14. Removing ball joints
15. Installing ball joints
16. Safe use of a strut compressor
17. Removing front or rear struts
18. Installing front or rear struts
19. Test driving a customer's car
20. Performing a grease job
21. Using a cutting torch
22. Using a mig welder
23. Installing copper welding wire on a MIG welder
24. Safe use of a wheel bearing puller
25. Removal and installation of headlights
26. Using the on-the-car lathe
27. Installing wheel bearings
28. Replacement of disc brakes
29. Using the brake washer on drum brakes
30. Removal and replacement of drum brakes
31. Safe use of a side-by-side lift
32. Putting away stock shipments
33. Maintenance of the oxygen generating machine
34. Replacing the oil in a compressor
35. Safe use of a fire extinguisher
36. Changing a serpentine fan belt
37. Removal and replacement of a master cylinder
38. The safe method of adding hydraulic fluid to an in-ground hoist
39. Safe use of a bench grinder
40. Changing a stone on a bench grinder
41. Changing spark plugs
42. Front end alignment
43. Performing a flush and fill on the engine coolant system
44. Installing a compressor for an air-conditioning unit

45. Adding refrigerant or other air-conditioning chemicals

46. Using the oil filter crusher

47. Removal and replacement of a brake rotor

48. Adjusting brakes

49. Removal and installation of heater hoses

50. Safe installation of a strut compressor

These lists should not be considered complete. Remember, as jobs are created or eliminated, the job list will change. Readers can probably assess the lists and add or delete jobs based on their own requirements. As a result of changes, the priority for selections of jobs may change somewhat.

A variety of jobs that would need a JHA in several industries and for several occupations have been presented here as a means of providing information on developing an analysis. There are a few key points on the job list and the examples used here that readers should keep in mind:

- The listings of the various jobs are typical for those occupations or industries. The reader can probably add many more to the list.

- In some situations, an organization may elect to use outside sources for assistance. As an example, tasks that are usually charged out to companies that repair forklifts, repair electrical problems, and perform facility repairs would be contacted for these services and the employees would not be involved.

- The extensive lists are reminders that an employee's job classification involves many more tasks and duties than may have been realized. JHA lists should be developed the same way—by using as broad a selection of potential jobs as possible.

- None of the jobs are listed in any form of priority. It would be up to management and the employees to designate which jobs should receive a priority for completion. Extensive lists, like those in this chapter, are usually the result of brainstorming on the part of workers and supervisors. Some may believe that they have identified every job in a facility, but other jobs will be recognized and added to the list. The master list is never complete.

- Organizations may wish to develop their master lists based on job classifications, like the example used for the power press operator. If there are 10 similar power presses and they essentially perform the same work, there is no need to complete a JHA on each machine. One JHA for all of these may be sufficient. The only exceptions to that statement are for those jobs that may be peculiar to certain machines but not to all of them. Many union contracts list all of the job classifications. Management can just record each classification and then have everyone add all of the various jobs for which each classification is responsible.

- Maintenance departments usually have the most extensive listings of jobs. This is a department that works independently of others, and the employees are often "all over the plant." This means that finding the time to complete a JHA on any job may take planning and coordination. In addition, maintenance workers are usually at a higher risk which makes the JHA all the more important. Management will have to recognize this need when the JHA program is being developed.

- There must be a coordinated effort to manage the master listing of jobs. If no one is paying attention to which jobs are chosen for JHA, there could be unnecessary duplication of work as well as the possibility that high priority jobs could go unselected. The purpose of JHA is to provide safe working procedures. Initially choosing jobs that are of a lower priority can be a disservice to the workers and to the company.

- The times in which we are living place a high premium on the clock. Every minute counts. Production is based on pieces per hour or the amount of footing or yardage being covered. If anyone and everyone in an organization is allowed to procrastinate on completing a job analysis, the quality of the program will suffer. Controlled job selection is very important. There should not be teams on each shift analyzing the same jobs. If the program is controlled, proper assignment of jobs will allow for each supervisor to complete their assigned JHA and will also allow him/her to review other JHA's completed in other departments or on other shifts.

Summary

A listing of all the jobs in a facility is necessary in a JHA program so as not to exclude any job from analysis. Do not rule out those jobs that appear to be less likely to cause an injury. It is recommended that supervisors and workers team up to develop the list as a precaution that no job is overlooked. Although there is no preferred way to organize the job list, all jobs should be assigned a priority based on the level of associated risks, and those jobs with the highest priority should be analyzed first.

Chapter 3

Completing the JHA Forms

Completing a JHA requires three steps: listing the basic steps of the job, identifying the associated hazards, and analyzing observations to make the job more productive and safe. This chapter discusses each step of this process and indicates where to include this information on the written JHA forms (Forms #2 and #3). The completion of any JHA should record and communicate to the worker how to do the job safely and correctly. Because completing JHAs becomes easier the more familiar the process becomes, the end of this chapter includes three examples of complete JHAs for common jobs.

Use Figures 3.1 and 3.2 as templates for the necessary JHA paperwork. Figure 3.1 depicts a common organization of a JHA rough draft (Form #2). Figure 3.2 depicts the final copy of a JHA (Form #3).

Standard Information

All JHA forms should include the following information:
- The JHA number from the job list
- The JHA title, which should be the same as listed on the job list.
- The person completing the JHA (On Form #2 the name can be handwritten, but on Form #3 the name should be typed. Signatures must appear on both.)
- Any employees who assisted (Provide the name and job title of the employee).
- The facility and department in which the JHA was completed
- The date of completion
- List of personal protective equipment required for the job

Basic Steps (Column 1)

There are no exact rules or guidelines for observing an employee at work. The idea is to observe and listen before asking questions or taking notes. The supervisor should inform the worker that they are about to complete a JHA on a particular task. Once the supervisor has observed the employee perform the task a few times, he/she is then ready to ask questions and begin preparing Form #2 (rough draft) of the JHA.

Job Hazard Analysis Form For Initial Draft
JHA Form #2

JHA NUMBER: _____ TITLE OF JHA: _____

LOCATION (FACILITY): _____

NAME OF PERSON COMPLETING JHA: _____SIGNATURE: _____

NAME OF PERSON(S) ASSISTING IN JHA: _____

DATE COMPLETED: _____ DATE REVISED: _____ DATE REVIEWED: _____

RECOMMENDED PPE: _____

Basic Job Steps	Hazards Present In Each Job Step	Correct And Safe Procedures For Completing The Job

Hazard Selection for the middle column: SB = Struck By; CW = Contact With; CBy = Contacted By; CB = Caught Between; SA = Struck Against; CI = Caught In; CO = Caught On; O = Overexertion or Repetitive Motion; FS = Fall At the Same Level; FB = Fall to Below; E = Exposure to Chemicals, Noise, etc.

Figure 3.1 JHA Form #2

Job Hazard Analysis – Approved Copy
JHA Form #3

JHA NUMBER: _____ TITLE OF JOB: _____ DATE JHA WAS COMPLETED: _____

PERSON COMPLETING JHA: _____ PERSON ASSISTING WITH JHA: _____

LOCATION / FACILITY: _____ DATE JHA WAS REVISED: _____

RECOMMENDED PPE: _____

Basic Job Steps	Hazards Present in Each Job Step	Correct and Safe Procedures For Completing the Job

Hazard Selection for the middle column: SB = Struck By; CW = Contact With; CBy = Contacted By; CB = Caught Between; SA = Struck Against; CI = Caught In; CO = Caught On; O = Overexertion or Repetitive Motion; FS = Fall At the Same Level; FB = Fall to Below; E = Exposure to Chemicals, Noise, etc.

Figure 3.2 JHA Form #3

The supervisor should begin by observing and writing down the basic steps of the task as they are being done. The supervisor should ask, "what is the first thing you do when you start this job?" The supervisor should indicate the response on Form #2 (see the discussion of forms in Chapter 2). Only a few words are necessary. Keep descriptions as brief as possible—usually 3 to 6 words. At this point the supervisor is only recording the "what" of the job, and should continue to ask, "What do you do now? Then what do you do?" until he has recorded all of the basic steps of the job.

Note: Most jobs will have about 10 steps in them. If the supervisor is at the midpoint of the JHA, and is at step 8 or 10, he should stop and reassess his efforts. The supervisor may be writing more information than necessary, or, the job is such that it requires more than one JHA. It is not possible to record every single movement someone performs while completing a job.

Recording too much information is a typical problem associated with individuals that are new to JHA. There are no hard and fast rules on how to identify the number of steps required for a job, but there is a guideline regarding the length of the JHA. Remember that a completed JHA is intended to make the employee's job more efficient and safe. If employees are handed JHA's containing 15, 20 or more steps, they will not be able to grasp the safety message. Learning will come much easier from JHAs that contain 10 or fewer steps. It is much easier to read and learn to do something that has brief, succinct points to remember rather than a lengthy document. If you want the worker to remember what they have just read, keep it brief.

In chapter one, the safe method of changing a flat tire was discussed, and a JHA for this job was created (see Figure 1.1). Note how the basic steps are recorded on this JHA. If the supervisor creating this JHA were attempting to record every movement a person made to change a tire, the written comments would probably look like the following:

1. Walk to car
2. Take keys out of pocket
3. Insert key in trunk of car
4. Turn key to the right
5. Reach into the trunk
6. Move materials in trunk out of the way to get to good tire
7. Unscrew tire from hold down
8. Lift tire from trunk
9. Set tire on pavement and roll to side of car
10. Lay tire down next to car
11. Return to trunk and lift out tools that are needed
12. Squat down and place pry bar behind wheel cover
13. Pry outward to pop cover off of wheel
14. Set cover to the side
15. Grasp scissors jack and place under car

16. Locate spot under car that has been provided for the jack

17. Place jack in special spot and snug up

18. Insert crank handle in jack

19. Rotate handle clockwise to slightly lift car

20. Grasp lug nut wrench

21. Insert wrench on lug nuts

22. Turn wrench counter clockwise

23. Remove lug nuts from wheel

24. Place lug nuts in wheel cover

25. Grasp handle of scissors jack

26. , 27, 28, etc., etc.

The example stopped at 25 steps and the flat tire has yet to be removed from the car. There is no need to break down a job in such fine detail. Imagine handing someone a document so large and asking him/her to follow it for safe job performance?

However, be careful not to go to the other extreme and underwrite. Some may write a JHA for changing a flat tire, listing the following basic steps:

1. Open trunk of car

2. Remove tire and tools

3. Remove flat tire

4. Replace with new tire

5. Place flat and tools in trunk

As the reader can see, this person left out many key details from the JHA. In situations such as this, where someone has omitted a lot of information, the question to ask is: "If you handed this JHA to someone, would they be able to safely complete the task?" If the answer is no, then there should be an effort to include the key steps of the JHA.

Key Hazards (Column 2)

Once the basic steps have been noted, the next step is to record the safety and health hazards that might be present. The best method of doing this is to question the worker concerning each of the basic steps. For example, "Joe, you said that the first step on this job was to unbolt the cover from the frame. What are the hazards or risks involved when you are doing this? How can you get hurt?" Once the comments and observations are made and the results are noted, move to the next step and ask, "How about this second step? Any hazards here?" Continue to follow this dialog through each step of the JHA until all of the steps have been covered.

If the worker is providing information on what he believes the hazards to be, and the supervisor agrees, this is a very good step toward achieving a quality product. The worker may point out hazards that the supervisor is not aware of and vice versa. However, if the worker omits some obvious hazards, the supervisor should question him further. Be specific about the nature of the hazard and indicate where in the job task it is found. For example, a supervisor might ask, "Do you consider this sharp edge on this metal sheet hazardous? I noticed that every time you cycled the machine you came very close to striking it with your elbow."

There can be many different reasons that the worker does not report hazards associated with the job. The worker may have accepted a condition thus that they no longer feel that it is dangerous. Or they may never have noticed the situation in the first place. Or they might realize that the situation is dangerous, but are still willing to face the hazard because they know that indicating the hazard could mean that they would have to change their method of doing the job or cause them to have to wear additional personal protective equipment (PPE).

There may be a difference of opinion on what the hazard is, but the supervisor should not walk away from an obvious hazard because they do not want to have a disgruntled worker. If the worker cannot prove that the hazard in question is not really a hazard, the supervisor should include such hazards on the JHA form.

The objective of the JHA is to identify and record the hazard and then develop a means of removing the hazard through guarding, distance, layout, PPE, engineering improvements, or work methods. If at the time of the initial observation the supervisor can correct a hazard, they should do the correcting at that time. Also, if the problem is something that needs corrective action, or a purchase order has to be prepared to correct the problem, complete these projects at that time. This is one way to show workers that management is sincere about correcting hazards and protecting them in the workplace.

If the hazards are such that major corrective action is required, stop the JHA, correct the problem, and then complete the analysis. The corrections may take several days or more. The worker should be told why the analysis was halted and when it will resume.

For each basic step of the job, there should be an indication of the number of hazards present. If there are eight basic steps to the job, then there should be eight hazard notations (one for each step). Each step of the job may have more than one hazard, but it is also possible for a step to be free of hazards. The supervisor may wish to place dashes across the column next to the number of the step or note that there is "No Hazard."

Analysis (Column 3)

Once the supervisor has created a complete list of basic steps and hazards of the job, it is necessary to provide relevant comments relating to each. Compare what is done by the worker in a basic step to what ought to be done. Is this step necessary to complete the task? Is there a way to alter the step to improve safety or productivity? It is essential to provide suggestions for correcting all hazards that were identified during the JHA? Discuss methods to prevent injury or illness from taking place. The analysis of the basic steps and hazards indicated requires more dialog regarding the who, where, why, when, and especially the how associated with getting the job done. The completed JHAs provided in the next section of this chapter provide good examples of the type of information to include as part of the analysis section of a JHA.

Additional analyses may arise as Form #2 is completed during the observation, and corrected (if necessary). Use this rough draft to produce the final written JHA (Form #3), which is the formal and controlled copy that should be properly typed and distributed in binders or mounted near the workstations.

Samples of Complete JHAs

The more JHAs you complete, the better you will become at recognizing the basic steps of a job, identifying hazards, and providing an analysis. Use the following three examples of common jobs found in many industries to become better acquainted with the JHA process. Provided are 29 sequenced photos of the three different tasks, which illustrate the step-by-step procedure for job evaluation. The purpose of the photos is to provide a means of assistance in the identification of the job steps. The photographs depict several steps of the jobs as examples of what supervisors might see while observing a worker. The written material identifies the basic steps of the job and their hazards. Finally, a complete JHA is provided for each job to demonstrate how the basic information gathered from the job observation should be organized on the final JHA form.

The three jobs selected for these examples are loading a trailer with product while using a forklift, locking out a switch in a circuit breaker box, and placing a pallet load of product on a rack. These three examples were chosen because they are commonplace and are a part of many processes in an industrial setting. Hundreds of thousands of docks are used each day to spot trailers for the purpose of loading and unloading product. In many other facilities there is a need to perform maintenance on machines that require lockout/tag out. Over 900,000 powered industrial trucks lift, move, and store product in facilities and at construction sites all across the U.S.

Loading a Trailer

Many workers perform the task of loading a trailer without really thinking about it because they have performed the same task so many times. As a result, many workers may not be aware of all of the hazards associated with this task. Injuries at the dock account for at least 10 percent of industrial claims. The key to keeping the dock safe is to allow workers to share in the process of developing safe work procedures.

The job of loading a trailer can be broken down into the following ten basic steps:

1. Have the trailer safety spotted at the dock.
2. Ensure that the trailer wheels have been chocked.
3. Inspect the landing wheels to ensure that they are free of defects, use trailer supports.
4. Open the trailer door.
5. Lift the dock plate chain.
6. Allow the dock plate lip to extend.
7. Walk the dock plate down onto the bed of the trailer.
8. Inspect the floor of the trailer.

9. Use the forklift to transport the pallets of product to the trailer.

10. Load the product into the trailer

Figures 3.3 to 3.11 illustrate the task of trailer loading.

Figure 3.3 Trailer backing up to a dock
Allow the trailer to back up to the dock door while standing clear, both inside and outside of the building.

Figure 3.4 Placing a wheel chock against a trailer tire
If necessary, go to the outside of the warehouse and chock the wheels of the trailer. If unsure as to whether the tractor operator chocked the wheels, check outside.

Figure 3.5 Checking the trailer landing wheels
While outside, check the condition of the landing wheels. Place supports under the nose of the trailer.

Figure 3.6 Employee opening trailer door
If the trailer door has to be opened in the warehouse, Inspect the door strap before pulling on it. Lift the roll up door with the legs and arms, not the back.

Figure 3.7 Pulling the dock plate chain
To lift the dock plate, brace an elbow on a knee and lift the pull chain with the other hand. This method will help in preventing back injuries.

Figure 3.8 Extending dock plate lip
Allow the dock plate to rise after pulling the chain ring. The dock plate lip should extend outward without any problems.

Figure 3.9 Walking the dock plate down
Walk the dock plate down and check to ensure that the dock lip extends well within the trailer bed.

Figure 3.10 Inspecting the trailer
Enter the trailer and inspect the floor and walls for holes or damage. Repair any damage to the floor before entering with a forklift.

Figure 3.11 Lifting the pallet and loading the trailer
Select the pallets or loads that have to be placed in the trailer. Follow safe driving rules while using the lift truck. Be alert for pedestrians. Load the trailer according to schedules and company requirements.

The hazards associated with this job are:

- Someone could be trapped between the trailer and dock as the trailer is backing. The trailer could also strike a person or object while backing. Even though the task of spotting a trailer is usually left to the driver, at times workers assist in the process.

- A worker could slip and fall on the stairs leading to the dock well or trailer spotting area. Workers also jump off of docks at empty dock doors to short cut the chocking of the wheels. Knees and ankles are usually injured during dock jumping.

- Once outside, the worker must place a wheel chock against the wheels of the trailer to prevent movement. Do not always rely on the driver of the rig to chock the wheels; hundreds of incidents take place each year when the trailer creeps away from the dock or when the trailer is prematurely pulled away from the dock. Another option in securing a trailer to a dock is through the use of a trailer restraint. Pushing a button at the dock door activates trailer restraints; the worker does not have to go outdoors to chock the wheels.

- While in the dock well the landing wheels on the trailer should be checked to ensure that there are no defects. Placing jack stands under the nose of the trailer will prevent tipping if the landing wheels collapse.

- Use care when opening trailer doors. Bend at the knees when pulling up a trailer door, the door could be binding in the track and difficult to move. On side-by-side doors, workers must stay clear when opening doors because product can easily spill out onto unsuspecting workers.

- It is now important to set the dock plate in place to allow the movement of a forklift into and out of the trailer. For manually operated dock plates reach down and grasp the pull ring. It would be easy to experience a back injury if the knees are not bent and the back is not flat. The chain is pulled and the dock plate will lift up and the lip of the plate will extend out. Walk on the dock plate to set it on the edge of the trailer.

- Inspect the trailer to prevent a lift truck wheel from going into a hole. If any holes are discovered they should be covered to prevent an incident.

- The forklift has to select the proper materials to load the trailer. In the process, pedestrians are vulnerable to the movements of the lift truck and the product being transported.

Figure 3.12 is the completed JHA on loading a trailer with a forklift based on the ten basic steps and their associated hazards indicated above.

Locking Out a Switch in a Circuit Breaker Box

Unfortunately, workers are seriously injured or killed each year from failing to properly lock and tag sources of power to equipment. A task that might take only a few moments to complete could be taken lightly because of familiarity of the job or other reasons. As a result, a form of energy, in this case electricity, can cause a serious injury. To keep this example simple, assume that an overhead light has burned out and an electrician believes that the ballast for the light must be replaced. The process of looking at an accurate listing of each switch allows the electrician to identify the specific box and switch and then move it to the "off" position.

Job Hazard Analysis—Approved Copy
JHA Form #3

JHA NUMBER: __5__ TITLE OF JOB: Loading an empty trailer with pallets of product. DATE JHA WAS COMPLETED: 12/14/00

PERSON COMPLETING JHA: John Belanger PERSON(S) ASSISTING IN JHA: D. Recci

LOCATION / FACILITY: East-Side Warehouse DATE JHA WAS RECEIVED: 12/21/00

RECOMMENDED PPE: Hard hat; safety shoes; gloves.

Basic Job Steps	Hazards Present In Each Job Step	Correct and Safe Procedures For Completing the Job
1. Ensure that trailer is correctly spotted.	CB – worker could be caught between backing trailer and dock. FB – worker could fall from the dock	1. Stay clear of the doorway while the trailer is being backed onto the dock. Keep others away from the area. Remove awareness chain or bar from the front of the dock door once the trailer is properly spotted.
2. Chock wheels; place jacks under trailer nose.	FS – on stairs going to dock well. SA – head could be struck against trailer. FS – on ice or snow	2. If the truck driver has not chocked the wheels of the trailer, go down the ramp/stairs to the dock well and chock the wheels. Use caution when walking on snow or ice. Hold onto hand rails; use ice- melt chemical if needed. When placing the chock, avoid bumping the head on the underside of the trailer. Place jacks under the nose of the trailer. If the dock is equipped with an automatic trailer restraint, push the button to activate the device.
3. Open trailer door.	O – while opening door. SB – falling product	3. If the trailer doors have not been opened, stand clear of the doors while unlocking and moving them. The best method is to stand behind a door and stay with it while swinging it open to avoid being struck by falling product. If it is a roll-up door, use your legs while the back is straight to lift the door. Be alert for falling material. Do not lift the door rapidly.
4. Extend dock plate.	O – while pulling on chain	4. Bend at the knees, grasp the pull ring on the dock plate and lift with the arms and legs. When the lip of the dock plate is fully extended, walk onto it to place the lip onto the rear of the trailer bed. Ensure that at least 3 or 4 inches of the plate are evenly placed on the trailer.
5. Inspect the trailer.	CW – wooden splinters. CI – worker could step in a hole	5. Turn on the moveable dock lights to provide lighting inside the trailer. Walk into the trailer to check the condition of the floor and sides. Be alert for holes in the floor. If necessary, place a wood or metal cover plate over a hole. If necessary, sweep the floor clean of debris. Wear gloves, and if sweeping, wear an approved dust mask.
6. Place product in trailer.	SA – forklifts can strike the sides of the trailer. FB – forklift and operator could fall from the dock if the trailer is not secured. SB – other workers or visitors could be struck by the forklift	6. Once the trailer has been prepared, use the forklift to move pallet loads into the trailer. Keep clear of the sides of the trailer to avoid causing damage. When going forward or backward with the forklift, look for pedestrians. Ensure the trailer is secured from movement before entering.

Hazard Selection for the middle column: SB = Struck By; CW = Contact With; CBy = Contacted By; CB = Caught Between; SA = Struck Against; CI = Caught In; CO = Caught On; O = Overexertion or Repetitive Motion; FS = Fall At the Same Level; FB = Fall to Below; E = Exposure to Chemicals, Noise, etc.

Figure 3.12 JHA on loading a trailer

The job of locking out a switch in a circuit breaker box can be broken down into the following nine basic steps:

1. Select the correct circuit breaker box.
2. Open the door to the box.
3. Select the correct switch from the listing on the door of the box.
4. Move the switch to the "off" position.
5. Place the proper lockout device on the switch.
6. Ensure that the device is correctly placed on the switch.
7. Add the lock to the device on the switch and include the tag.
8. Correctly fill out the tag.
9. Remove the key and put it in your pocket.

Figures 3.13 to 3.19 illustrate the circuit breaker box lockout procedure.

Figure 3.13 Circuit breaker box
Check the written lock out program to determine which circuit breaker box must be locked and tagged. In this case, it is box 107.

Figure 3.14 Opening the circuit breaker box
Open the box to initiate lockout / tagout.

Figure 3.15 Turning off a switch
Check the master listing of what each switch controls on the inside door of the box. Locate the correct switch and move it to the "off" position.

Figure 3.16 Installing a lockout device
Place a lockout device on the switch while it is in the off position. Ensure that the lockout device is secured on the switch.

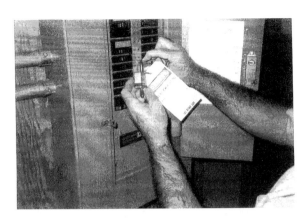

Figure 3.17 Locking out the switch
Slide a lockout tag over the lock and insert the lock in the hole provided on the lockout device.

Figure 3.18 Completing the lockout form
Complete the tag with "who" put on the lock on, "why" was it placed there and "when" was it placed there.

Figure 3.19 Removing the key from the lock
Check to ensure every thing is correctly in place and that it is safe to replace the ballast on the light fixture. Place the key in your pocket so no one can remove the lock on the circuit breaker box.

The hazards involved in exposures to electrical energy are definitely life threatening. Every facility must have a written program in regards to the control of different forms of energy because OSHA requires employers to have such a program. The written hazardous energy control program would identify the various pieces of equipment and the means of obtaining a zero energy state. The photographs illustrate circuit breaker box number 107. In the written plan, each circuit breaker box would be appropriately numbered and the contents properly identified. This procedure can only be done one way—to check each individual switch and make sure that what has been identified as the function of the switch is accurate. Failure to shut off and lock out the correct switch can be very hazardous.

In this example, the hazards are:

- As mentioned above, ensure that the correct switch has been selected and moved to the "off" position.
- Properly place the lock on the device being placed on the switch. It should be noted that this job step is not dangerous in itself; it is the consequences of the failure to eliminate the electrical energy that are the real hazard.
- Unless there are missing switches in the control panel box that pose an electrical shock hazard, locking and tagging a circuit breaker box is a relatively safe process. However, the electrician is putting his life on the line if the job is not done correctly. This is why a job safety analysis is so important for each job in the facility.

Figure 3.20 is the completed JHA on locking out a switch in a circuit breaker box, based on the nine basic steps and their associated hazards indicated above.

Job Hazard Analysis—Approved Copy
JHA Form #3

JHA NUMBER: ___21___ TITLE OF JOB: ___Locking out a switch on a circuit breaker box.___ DATE JHA WAS COMPLETED: ___12/27/00___

PERSON COMPLETING JHA: ___Ken Carlton___ PERSON(S) ASSISTING IN JHA: ___Peter Reis___

LOCATION / FACILITY: ___Warehouse – East Side___ DATE JHA WAS RECEIVED: ___12/15/00___

RECOMMENDED PPE: ___Safety Glasses___

Basic Job Steps	Hazards Present in Each Job Step	Correct and Safe Procedures For Completing the Job
1. Select correct box.	1. No Hazard	1. Check the written lockout / tagout program to obtain the information for the exact circuit breaker box and switch to use.
2. Open the door.	2. SB – Movement of lift trucks	2. Go to the correct circuit breaker box and open the door. Select the correct switch that will have to be turned off. Stay in marked walkways going to the circuit breaker box and be alert for the movement of lift trucks.
3. Turn off switch and add device and lock.	3. No Hazard	3. Once the correct switch has been identified from the listing on the box, move the switch to the "off" position. Place the individual lockout device on the switch. Once in place, add the lock and lock it onto the device.
4. Fill out the tag.	4. No Hazard	4. Complete the necessary information on the tag: the date, who placed it on the switch, and why. Remove the key and place it in your pocket.

Hazard Selection for the middle column: SB = Struck By; CW = Contact With; CBy = Contacted By; CB = Caught Between; SA = Struck Against; CI = Caught In; CO = Caught On; O = Overexertion or Repetitive Motion; FS = Fall At the Same Level; FB = Fall to Below; E = Exposure to Chemicals, Noise, etc.

Figure 3.20 JHA on locking out a circuit breaker box

Moving and Storing Product on a Rack

The job of moving and storing product on a rack can be broken down into the following nine basic steps:

1. The operator or a fellow worker should inspect the load to ensure that it is acceptable for handling and storage.
2. The operator should drive under the load, making sure that the forks are in far enough to safely handle the load.
3. Drive the load to the designated storage spot. Be aware of pedestrians and other obstacles while driving. Keep the load as low as possible.
4. After aligning the lift truck to the required storage area, begin lifting the load. Be sure the load is clear of the racking.
5. Safely position the pallet on the racking. Do not allow anyone near the load or lift truck while it is raising or lowering the load.
6. Before backing out with the forks, check behind and around you.
7. Lower the forks to within 2-4 inches of the floor for safe clearance.
8. Look before backing out of the spot.
9. Safely drive to the next task or repeat the same steps.

See Figures 3.21 to 3.28 for the correct procedures for loading a pallet on a rack.

Figure 3.21 Checking the pallet load
First inspect the pallet and load.

Figure 3.22 Driving the pallet to the storage area
Safely drive to the rack storage area.

Figure 3.23 Preparing to lift the pallet
Properly position the lift truck for the correct lifting and placement of the load.

Figure 3.24 Raising the pallet
Raise the load and observe the movement to avoid striking other product.

Figure 3.25, Placing the pallet on the rack
Extend the scissors mechanism and place the pallet on the rack.

Figure 3.26 Removing the forks from the pallet
Retract the forks toward the mast and back up after looking to both sides and behind the lift truck.

Figure 3.27 Lowering the forks
Lower the forks to within 2 to 4 inches of the floor while observing the lowering of the forks for clearance.

Figure 3.28 Leaving the storage area
Look behind before moving the lift truck, and safely proceed to the next load or task.

Even though this task appears to be very simple, it requires a great deal of skill and there are several hazards associated with the steps.

- If loads and pallets are not inspected while on the floor, there could be a problem when the load is in the air or in storage. Note that the load in the photo is shrink-wrapped. Shrink-wrapped loads are safer because the plastic wrap helps to prevent product from falling to the floor. There can be serious injuries to anyone that is struck by falling loads. Also, inspect the pallet while the load is on the floor. In the event that the pallet is defective, it can be removed and replaced with a good one. Defective pallets can fail while in storage and product could fall on workers.

- Use the forks to enter the pallet for transport. Once the forks are placed evenly in the pallet, lift slightly and then tilt the load. Always look for pedestrians, other pieces of power equipment, and obstacles while driving. Keep the load as low as possible for safe floor clearance.

- Drive to the designated storage area. This is where the skills of the operator come into play. The forklift in the photos is a narrow aisle reach truck and it takes a great deal of skill to operate it—much more than a sit-down model. The operator must align the machine with the storage area so that there is a smooth transition in the placement of the load and removal of the forks.

- Lift the load and observe it while it is being raised. The lift truck must be back far enough so that the load being lifted does not strike other loads. Always keep others from the area so they are clear of loads and the power equipment.

- Once the load has reached the proper height, pull forward and extend the load with the scissors. When sure that the load is correctly over the storage spot, tilt downward and lower the forks. Place the pallet on the racking and ensure that the pallet base is located evenly on the front and rear sections of the rack.

- Retract the scissors mechanism toward the mast and back up. Always look behind and to the sides before backing.

- Lower the forks and observe the forks to ensure that there is the correct clearance. Lower the forks to within 2-4 inches of the floor for safe floor clearance. Never drive with elevated loads or forks.

- Look to both sides of the lift truck before turning. This particular model of lift truck has an extremely sharp turning radius.

- Once the lift truck is out of the aisle, proceed to the next load or task. Always be aware of pedestrians and other lift trucks.

Figure 3.29 is the completed JHA on moving and storing product on a rack based on the nine basic steps and their associated hazards indicated above.

Common Errors

Accurately completing the JHA forms can be a tricky business. Important information can be inadvertently omitted or incorrectly recorded. Figure 3.30 uses a standard JHA form to depict the most common errors made when completing the written portion of the JHA.

Summary

Interviewing workers and creating JHA documentation may seem like an overwhelming task. At first, the process of working with an employee and asking questions while recording his or her responses may seem difficult. In time, the practice of focusing on the correct methods used to gather important information during the observation will improve and the time and effort of the program will be rewarding to everyone.

It is important to follow the correct procedures for completing the written portions of a JHA when performing an observation. Remember to keep comments brief and only identify the "what" of the job when recording basic steps. Listen to the worker and observe his routine to identify the hazards associated with the job. Finally, provide a thorough explanation of exactly how to do the job properly and correct the hazards that exist.

Job Hazard Analysis—Approved Copy
JHA Form #3

JHA NUMBER: __54__   TITLE OF JOB: Placing pallet loads on racking or in storage.   DATE JHA WAS COMPLETED: 2/15/01

PERSON COMPLETING JHA: Sam Spitale   PERSON(S) ASSISTING IN JHA: C. Hazel

LOCATION / FACILITY: East-Side Warehouse   DATE JHA WAS RECEIVED: 1/28/01

RECOMMENDED PPE: Hard hat; steel toe shoes; gloves; safety glasses.

Basic Job Steps	Hazards Present In Each Job Step	Correct and Safe Procedures For Completing the Job
1. Inspect pallet loads.	1. SB – forklifts in the area CW – splinters from pallets	1. Inspect all pallet loads before lifting them. Check to ensure that shrink-wrapping is in place. Look at the condition of the pallet to ensure there aren't any defective boards. If defective boards are present, replace the pallet. Be alert for passing forklifts; keep clear of their movement. Wear gloves to prevent splinters.
2. Lift the load.	2. SA – load against other objects	2. Ensure that forks are wide enough and pull forward with the forks and enter the pallet. Look at the work area to keep from striking anything when the pallet is lifted. Keep the load as close to the floor as possible. If the load is too high to see over, operate in reverse.
3. Drive to the designated area.	3. SB – workers or objects struck by forklift SA – load against other objects	3. Drive the load to the spot where it will be deposited. Be alert for other workers, forklifts and objects in your path of travel. Sound your horn where necessary to alert others and follow safe driving rules.
4. Position the forklift and lift the load.	4. SA – load against other objects	4. Pull forward into the area where the load will be placed on the racking. Be aware of everyone around you and keep from striking objects with the load or forklift.
5. Place the load on the racking.	5. SB – falling objects SA – load against other objects	5. When you are sure you are in the correct alignment for the racking, lift the load and place it on the racking. Keep back far enough from the racking to keep from striking it. Do not allow anyone to walk under the load. Pull forward cautiously and place the pallet on the racking. Be sure the front and rear of the pallet are secure. Tilt forward and lower the forks. Be aware of sprinkler heads or natural gas pipes.
6. Lower the forks and pull away.	6. SA – forks or forklifts against objects SB – workers or objects struck by forklift	6. Look to the sides and behind you before backing up. Be sure the forks are clear of the racking area before lowering. Safely drive to next task.

Hazard Selection for the middle column: SB = Struck By; CW = Contact With; CBy = Contacted By; CB = Caught Between; SA = Struck Against; CI = Caught In; CO = Caught On; O = Overexertion or Repetitive Motion; FS = Fall At the Same Level; FB = Fall to Below; E = Exposure to Chemicals, Noise, etc.

Figure 3.29 JHA on storing product on a rack

Typical Errors on JHA Forms

JHA NUMBER: _____Omitted or Incorrect_____   TITLE OF JOB: _____Omitted or Incorrect_____   DATE JHA WAS COMPLETED: _____Omitted_____

PERSON COMPLETING JHA: _____Omitted or Incorrect_____   PERSON(S) ASSISTING IN JHA: _____Omitted_____

LOCATION / FACILITY: _____Omitted or Incorrect_____   DATE JHA WAS RECEIVED: _____Omitted_____

RECOMMENDED PPE: _____Omitted, Incorrect, Does Not Match Descriptions Below_____

Basic Job Steps	Hazards Present in Each Job Step	Correct and Safe Procedures for Completing the Job
• Too many words are used in this section (6 words should be the maximum). • Words do not describe the basic job step. • Basic job steps go beyond identifying just the "what" in the JHA.	• Hazards are not identified by code (SB, CI, etc.) • Real hazards are not recognized and recorded. • Hazards are overstated; in some cases the hazards do not exist.	• The safe procedure does not identify the information in the basic step and / or does not identify the hazards. • Solutions to preventing the hazards are not provided. • Alignment of numbered sections / steps are not in a single horizontal row. • Narrative is difficult to understand. • Specific PPE that is required has been omitted from the explanation in a step. • Too little narrative, the job step and job hazards are not identified. • Too much narrative, overwriting, paragraphs are too long.

Figure 3.30 Common errors on JHAs

Chapter 4

Performing a Job Observation and Correcting Hazards

One of the most difficult tasks of completing a JHA is identifying and reporting the hazards associated with a job. Although it is impossible to predict all of the possible risks, completing a JHA can uncover a significant number of workplace hazards, the correction of which will significantly improve workplace safety. During a JHA, the supervisor has a perfect opportunity to identify, report, and correct existing hazards because the supervisor is positioned at the workstation or at the particular worksite where the task is performed. However, the task of looking for and correcting hazards is not limited to the supervisor. Workers have a significant role in making the workplace safer because they are the ones who regularly perform the task. Observing and questioning workers can reveal much about a job. This chapter will discuss how to identify and correct hazards through job observations.

Continual Observation

Performing a planned or unplanned facility walk-through can reveal much about worker safety. Observing workers at their workstations in regard to the JHA process should be a planned process, but it does not always have to be a formal observation that utilizes a form to document the observation. The job safety observation (JSO) is one of the better methods to correct unsafe conditions or behavior during plant inspections or a facility walk-through. The JSO is a strong component of any safety program.

Management has an obligation to keep its eyes and ears open during every tour of the plant. If something is wrong or unsafe, it should be corrected when it is noticed. Do not wait for a formal JHA observation. Every time a manager tours the facility, they should be sure to give an employee a positive comment or to ask if there is anything they can do to improve the employee's working conditions.

There are hundreds of items that deserve a look while walking through a plant. When a supervisor goes out onto the factory floor or the jobsite, there are many items that he should be aware of, such as:

- Is the production proceeding on time?
- Are the parts being produced as required?
- Is the quality of the product or task meeting standards?
- Are the conditions that the worker is exposed to interfering with safety or productivity?
- Can other workers or adjacent processes pose a risk to the worker?
- Has the worker that is being observed just returned from being away from work as a result of an illness or injury?
- Is the employee new to the job?

Obviously, no one can find and correct every single item. That is why supervisors should take a different direction during each walk-through and focus on the obvious hazards while inquiring about the not so obvious. The worker knows the process they are performing better than anyone else, and can therefore be a vital resource to the supervisor. After all, who knows more about their job and surroundings than the forklift operator, punch press operator, carpenter, or laborer?

Interacting with workers during site visits is very important. Consider the following example: A plant manager in a mid-west manufacturing plant was never successful in achieving good safety performance, cordial employee relations, fewer grievances, and increased productivity. He constantly announced "his open door policy" for any worker, but he never had a single worker take advantage of it. His talent was in the working knowledge of the machines and work processes. The plant manager would periodically be called to the shop floor to troubleshoot a defective machine or process and usually succeeded in getting the machine to run properly. He rarely made eye contact with any worker; also, he failed to greet workers on the way to and from the machine. Is it any wonder that workers did not trust the plant manager? Is it any wonder that there were hundreds of grievances each year on a variety of issues? How much could have been accomplished if the welfare of the worker was uppermost in the mind of the plant manager?

Supervisors usually adopt the behavior of the plant manager or owner. If the plant manager does not endorse plant safety and recognize the value of the worker, there will probably be problems with productivity and safety. If there is to be improvement in the working conditions of the work site, then every member of management has the duty to foster improved employee relations. Much of this improvement should focus on safety conditions. Most of the visual checks during a shop tour only take a few moments and they can be very beneficial to everyone concerned. Workers are likely to have more respect for their supervisor and the company they are working for if they feel that they have a voice and that management has a concern for their safety. Casual and incidental looking during site tours on the part of the supervisor can uncover many unsafe conditions and practices. The supervisor should be able to spot unsafe conditions and unsafe worker practices because they have honed their skills to recognize these things. The key to effective incidental safety observations is to look at what workers are doing as the supervisor goes from place to place. Every member of management should look around the department with safety on his mind. Sometimes, supervisors have to be reminded of the importance of taking safety personally and seriously. A safety director was heard to always ask supervisors, "What did you do for safety today?" At first, the supervisors thought that the safety director was paying them a slight insult, until it dawned on them that their job was to resolve safety problems and improve working conditions—not just to produce parts and meet production schedules.

Before the JHA

Supervisors are usually in the best position to conduct observations. They know their department, they know the workers, it is their worker that may be injured, and it is their production that will suffer as a result of the loss. As a result of the worker-supervisor relationship, the person that can achieve the most in the observation process is usually the worker's immediate supervisor.

Before beginning a JHA, you must choose a job to analyze and a worker to observe at that job. This choice will influence the hazards supervisors will be able to identify. There are several items to consider when the supervisor is making a determination of who will participate in the JHA and what JHA will be selected from the master list. Of course, the specific job has already been selected from the master list. The supervisor should collaborate with the worker on what job should be selected.

It is not always best to choose to observe workers who have been with the company the longest. Even though the worker may have plant seniority, he may not be experienced enough on a particular job to assist in completing a JHA. Also, some workers may have the most experience in terms of years with a particular job, but have not performed that job for a while and so they may not know the details of any changes to the process that may have occurred since they last performed the job.

Of course, do not choose the worker with the least experience, because they have not been on the job long enough to help supervisors identify the basic steps of the job or to identify hazards. Indeed, they might still be learning the process themselves. It is also good to question whether the worker being considered for the JHA has a history of unsafe behavior or multiple injuries? Workers with this type of history may be repeatedly injuring themselves because they are not familiar with the basic steps of the job.

Remember that the supervisor is counting on the worker to offer specific details on how to complete the task. The best worker to observe is one who is not new to the job and has performed the job for a few months immediately preceding the JHA.

During the JHA

Carefully observe the entire workstation and worker behavior while completing a JHA. Consider the presence and habitual use of the physical items at, on, or near the job being observed. All supervisors should be skilled enough to recognize these typical workplace conditions while discussing the job and recording the JHA with the cooperation of the employee. Observation of the worker performing the task is essential to the JHA process. The movements and steps that workers take must be analyzed to ensure that they won't injure themselves or others.

To determine the hazards of the work practices of the employee, the supervisor should question if the worker can:

- Be caught between two or more objects?
- Be contacted by hazardous chemicals or objects that are too hot or cold?
- Be exposed to airborne hazards such as welding fumes, vapors, gases, mists, dusts, radiation, or noise?
- Be injured by taking unusual postures or positions?
- Be struck by a moving object?
- Be caught in a hole or other entrapment to the body?
- Make contact with hazardous energy?
- Fall to a lower level?
- Slip, trip or fall on the same level?
- Be caught on a moving object?
- Strike against a fixed or moving object?

When observing the workstation, typical items that involve physical hazards include:

- Housekeeping practices.
- The use of personal protective equipment.
- The safe storage and use of chemicals.
- Lighting.
- Machine guarding.
- Ergonomic factors.
- Noise, radiation or any other environmental hazard.
- Conditions associated with the floor, hand railing, ladders, and stairs.
- Extremes of cold and heat and protective clothing.
- Conditions associated with powered equipment.

Analyzing the conditions of the workstation and the worker's behavior will be discussed in greater detail in Chapters 5 and 6.

Correcting Hazards

Surely, the JHA will uncover some physical hazards that must be corrected. Hazards that can be corrected immediately should be. Unsafe conditions that could be corrected during the JHA would include such items as: re-installing a machine guard, moving a pallet which is covering a walkway line, issuing a pair of safety glasses, replacing defective lighting and placing wheel chocks against a trailer wheel. Other hazards may require maintenance department or outside contractors to be contacted to build a guard, install a handrail, conduct air sampling, improve the lighting, etc. It is very important to correct these obvious hazards as soon as possible because they could easily result in an immediate injury.

A safe workplace is management's responsibility, and it is expected that management correct all of the identified workplace hazards within a reasonable amount of time. Once discovered, every hazard should be neutralized so losses are not allowed to occur. Correcting hazards that carry a greater threat (either a greater chance of serious injury or more frequent injury) should take priority over other hazards. A timetable for correcting all hazards should be created. Management has a duty to control costs, and it is less expensive to correct the hazard than to risk an injury or fatality.

If supervisors wish to convince workers that the company is serious about their efforts in the safety program, the completion and resolution of maintenance-related safety hazards must be very convincing. When a worker is willing to come forward and report an unsafe condition or a defective machine, it is management's duty to ensure that the items are corrected. On the other hand, collecting work orders along with comments from the workers and failing to correct the hazards will not help the safety program or employee relations.

Before creating any written material for a JHA, including a rough draft, the supervisor should get a "feel" for what is going on. There are times that the job will require corrections to the job or work area before the actual JHA is written. It is the astute supervisor that will stop the JHA process and tell the worker that he must have certain items corrected before the supervisor can continue. The corrections can be made in a relatively short time or maintenance will have to be summoned. The time that maintenance can complete their work on the problems may take several hours, days, or weeks. It makes no sense to take the time to complete a JHA if the worker is working near obvious hazards and they cannot be corrected. As with any workplace hazard, they should be corrected when they are discovered. Note the example below that demonstrates the benefits involved in the delay of the JHA.

At a facility in the southeastern U.S., a supervisor that was relatively new to the organization was completing his first JHA with a worker whose occupation required him to select product, place it on carts, wheel the parts to a large table, and prepare the product for shipment via UPS. It was a tough demanding job that kept the worker busy for the entire shift. In fact, there should have been two workers at this job classification, but that was going to take place several months into the future. What was important to both the supervisor and "UPS man" was to make the job safer now, not wait until another worker showed up in the future.

The supervisor studied the various steps and tasks with which the worker was involved, and settled on the selection of "Preparation and shipment of the UPS order" as his job to analyze. The supervisor, with clipboard in hand, observed the worker for several minutes to get a feel for how the job was accomplished. Being that the selected orders had to be processed as quickly as possible, the supervisor studied and listened to the worker and tried not to interrupt the process while making just a few comments on his notepad. Based on what he observed, the supervisor did not proceed to complete his JHA because he felt that there were too many things wrong with the job that needed fixing right away. So he informed the worker that he would delay the analysis until the immediate problems were corrected.

At the end of the shift the worker went home. The supervisor then proceeded to make some dramatic changes to the work area. He had told the worker that he first wanted to observe the job and ask questions before writing his JHA. What the supervisor noted was that the worker was working harder than he had to. During that evening, he made the following changes to the work area:

- He raised the worktable six inches so the worker would not have to bend forward to do the job. This was accomplished by placing large wooden blocks under the table legs. He guessed at the proper height, but knew that the height could be adjusted if necessary.

- He placed a large spongy floor mat in front of the worktable. The worker was standing on a low profile wooden pallet at the time of the observation. The floor mat was being used at the front of an emergency exit for wiping the soles of the shoes. The supervisor assumed that once someone goes out of an emergency exit, it is not likely that they will enter by the same door.

- He moved an out-of- place overhead fluorescent light directly over the top and a little to the rear of the table. The light did not have fixed wiring but was powered by an extension cord. The supervisor knew that once that light was properly placed, fixed wiring would be needed. He also cleaned the reflective shield and wiped off the bulbs to create more light.

- He then arranged the small picking containers and tape rolls and placed them on the sides of the table so the worker would not have to reach or bend forward for the parts.

- The scale for weighing the product was raised to a higher level to make placing and removal of the parts easier.

- He placed a section of a rubber sheet on the edge of the table to make it more comfortable for the worker if he leaned against the table. The narrow sheet was held in place with duct tape and did not look "tacky."

With a little effort, the supervisor had turned an ugly little corner of the plant into a safer area to work. His observation of the employee doing the job, along with employee comments and a review of past injuries at this work area prompted him to improve the working conditions. Now he was ready to complete a formal JHA with the "UPS man". Needless to say, the worker went home from the job each day less tired than in the past. In addition, the supervisor eliminated job hazards for just an hour's work and for very few dollars spent in the process.

The story above is true. The supervisor was curious as to why the job was allowed to be performed as it was. He asked several other supervisors why those job conditions were allowed to go uncorrected, despite the obvious ergonomic problems. He was told that the plant manager only asked the supervisors to focus on the hazardous jobs for analysis, not "regular jobs." The example makes several points. First, if the supervisor can eliminate a hazard while conducting the JHA, he should do so. It makes no sense to write a JHA and prepare a long list of items to correct if the correction can be done at that time. Once the initial problems are corrected, then the supervisor may proceed with the analysis. In cases such as these, a delay that involves safety improvements is advisable. The other point is in regard to underestimating the need to include all jobs in the JHA program. Yes, some jobs are hazardous and deserve to be evaluated, but sometimes the jobs that appear to be unseen, unknown, and unobserved must be included in the process.

Summary

It is very important, in any safety program, to ensure that those conditions are corrected that can cause harm to workers. Both conditions and behavior are to be considered in every injury investigation, facility inspection, job safety observation, and job hazard analysis. Supervisors should focus on both unsafe acts and unsafe conditions. Correcting and identifying hazards should not only be the domain of formal JHA observations; during any of the visits that a supervisor makes through his department he or she should make every effort to keep the department safe.

While observing the worker and the job being performed, it is essential for the supervisor to point out and record unsafe conditions. Where possible, the conditions should be corrected at that time. Many conditions can be corrected rather easily while others may require maintenance work requests or a capital expenditure. The maintenance system that allows scheduling of any item that needs correcting in the plant must also be designed to include safety related items.

Chapter 5

Identifying Workplace Hazards

Preventing injuries and illnesses in the workplace is a never-ending job. Through job hazard analysis, supervisors can focus on each task and the hazards associated with it to identify most of the potential causes of injuries. Unsafe conditions and behaviors will only be eliminated after workers and management identify what they are and why they exist.

There are many causes of injuries, which makes prevention difficult. The simplest of injuries may have three or more causes. When supervisors conduct injury investigations, they usually identify only a single cause. It is the responsibility of management to search for all of the causes of a particular injury and eliminate them so that there is little possibility of a similar injury occurring.

Unfortunately, the true causes of injuries are often undiscovered because the worker is thought to be the only problem. Earlier studies in safety argue that almost ¾ of all injuries are the result of unsafe worker behavior. Although it may be true that the worker is involved in a high percentage of injuries, it is doubtful that the employee is solely responsible for the injury. Proper job hazard analysis may reveal that processes and work conditions share responsibility for workplace injuries. Management must also take some of the responsibility because injuries become more prevalent when safety programs are not established or supported. (Appendix A provides more information on how to develop comprehensive safety and health programs.)

Workplace conditions and the behavior of both management and labor have to be considered when investigating injuries. Yes, workers do contribute to injuries by injuring themselves or others through improper work habits or procedures. But it is important to look at the reasons why workers do not always work safely and why management is not able to control the hazardous conditions in the workplace. Safe workplace conditions are more than just the placement of guards on machines, safety glasses, hard hats, and clean floors. The presence of safe or unsafe working conditions and behavior is directly related to the values that management establishes. Workers will more likely abide by workplace rules if they are proper, equally enforced, and fair, and if there is professional training to support the rules.

59

Unsafe Conditions

Unsafe working conditions play a major role in the causes of injuries. The unwillingness or inability of management to plan properly, correct unsafe conditions, provide ongoing maintenance, or purchase safer machines can easily result in worker injuries. At times, one will hear the statement that despite the broken part, defect or hazard, a smart worker can work around the hazard and not be injured. This is true to a degree, but how long can this belief last before someone falls victim to the hazardous condition? Will a new worker know of the hazard that he is supposed to " work around"? Is the defective part also affecting productivity? Is there a potential for an OSHA citation?

When unsafe conditions are detected—including during the JHA process—management should correct them as soon as possible. In some cases, such as if there is an open hole, exposed electrical wiring, or a chemical spill, time is of the essence. Any delay in correcting these situations can cause serious injury or death. In some cases, a part may have to be ordered. In the interim, until the part arrives and the condition is corrected, precautions such as lock out, signage, lights, warnings, and barricades are necessary.

In any safety program, there is a need to establish procedures for recognizing and correcting unsafe conditions in the workplace. An unsafe condition can involve items such as protective equipment, machines, air and noise contamination, tools, chemicals, workplace materials, product, and the facility or worksite. Injuries and illnesses are more likely to occur if unsafe conditions are present. A high percentage of OSHA citations are issued for unsafe conditions. Many of the American National Standards Institute (ANSI) and National Fire Protection Association (NFPA) standards also focus on establishing safe conditions in the workplace. JHAs provide an excellent opportunity to identify unsafe conditions because the supervisor will be in the workplace at the worker's machine or operation.

Unsafe conditions can be a part of the design of a machine or operation. Before a machine is placed in the workplace, it should be evaluated to determine, among other things, if it will have all of the appropriate guards, generate noise levels below 85 decibels, contain electronic safeguards, and be free from ergonomic hazards. If more attention were given to the safety through design principle, machines and operations would have fewer hazardous conditions. Management must consider the safe design of equipment used in the workplace. Perhaps the weight and lifting capacity of a piece of powered equipment is adequate for the product being handled at the time of purchase. However, as product changes and physical demands of the plant change, the equipment may be too light to perform the job.

The following forklift example demonstrates the common safety problem that results from improperly choosing equipment. An employer will purchase a certain capacity of lift truck to accommodate the product being handled at that time as well as economizing on the purchase of a lighter capacity machine. As time goes by and the demands of the job change, the lift truck is required to handle heavier product. To make up for the lighter capacity, the employer has a few workers sit on the counterweight to keep it from rising up when a heavy load is lifted. The added weight to the counterweight provides some added traction for the rear wheels but also exposes workers to potential injury and an OSHA citation.

The effects of time on equipment, and the quality of the equipment's construction must also be considered as contributors to unsafe conditions. Building products such as wood, concrete, metal and stone do not last forever. Wooden stairs and railings can easily dry rot over time and create unsafe conditions. Floors can be made of concrete and wood and yet collapse because of loading conditions and the effects of time. For example, a forklift was moving air conditioning units on the fourth floor of a multi-story warehouse when the floor collapsed without warning. The forklift fell several floors through a hole that suddenly appeared in the concrete floor. The fall was 40 feet and the forklift came to rest and was buried with concrete and steel. The operator survived because he was protected by the overhead guard and was wearing a seat belt. He only suffered a minor fracture to a leg but could have easily been killed.

Placement of product is also a common safety issue in the workplace. In warehouses, distribution centers, and retail stores, the business requires a constant shifting, handling, and moving of product. In some cases, the stacking and storage of the product is not safe. The loads can topple over unexpectedly or collapse on their own, or they could be bumped and collapse. There are many workplace incidents involving falling product that are caused by being bumped or struck by a powered industrial truck. A forklift can easily strike a stack of product that is not secured properly and cause it to fall on a worker. In some cases, a forklift placing a pallet load of product on a storage rack in one aisle can easily push another pallet off of the rack in the adjacent aisle. Anyone walking in the aisle could easily be struck by falling product. Employers should set limits on how high product is allowed to be stacked without the benefit of racking or shelving and set standards on how the product is to be handled and placed to make it secure.

Where the floor layout and design and safety of workers in a plant has not been given much thought, there can easily be situations that threaten the life of workers on a daily basis. These unsafe conditions can pose a risk because of the placement of the worker near the operation. If a worker is facing a machine and has his back to an aisle way that is used by powered industrial trucks, these passing machines can easily strike the worker. Sometimes the lift trucks pass within inches of the worker. It is necessary in these circumstances to erect barriers to provide protection to the worker or reposition the machine.

Plant design and layout for foot and vehicle traffic can present a serious hazard. If workers must cross busy intersections, they will surely be endangered by vehicle traffic. The erection of an elevated walkway above the intersection would provide safe passage to pedestrians. Stairs and hand railing over the top of conveyors provide this same protection. If management does not want workers to climb over or under a conveyor, there is a need to place safe walkways in the facility at key locations. Otherwise, workers can easily climb onto or over moving or stationary conveyors, and they might injure themselves.

The placement of locker rooms is very important. Many engineers design locker rooms in the center of a plant for convenience. However, in order to reach the locker rooms located in the center of the plant, workers must walk into the facility without their personal protective equipment and pass dangerous machines and operations. Workers entering the plant do not have immediate access to equipment such as safety glasses, hard hats, steel toe shoes, respirators and gloves as they move through the part of the plant where this protective equipment may be necessary. At the end of the shift, this process is repeated. Many injuries have occurred to workers while they were walking to and from their workstations or the locker room. Many of these workers were on company property but not on the time clock. Obviously, the best location for a locker room is adjacent to the parking lot. This location allows workers to enter the locker room as soon as they enter the building, and wear their equipment to and from their jobs safely.

Conditions can also be affected by inadequate or poor maintenance procedures. The failure of wire rope, pulleys, bearings, electronic limit switches, guards, and safety features on vehicles can be traced back to unsafe conditions. A load falls from an overhead crane and narrowly misses a worker. The main cause of the incident was a frayed wire rope that should have been inspected and replaced weeks ago. An employee loses the tip of a finger in a punch press while removing a part from the die. The main cause of the injury was the failure of the lights on an electronic guard. He placed his hand in the area of the dies to remove the part and assumed that the machine would not cycle. Had the "light guards" been working, the machine could not have cycled. This was a problem on the earlier shift but nothing was done about it. At a construction site a laborer was struck by a piece of backing power equipment. The main cause of the injury was a non-functioning back-up alarm. He was not aware of the backing machine, and therefore was not warned to move out of the way.

Supervisors, plant managers, safety committee representatives, and workers should be aware of common unsafe conditions, including the following:

1. Lack of or inadequate safeguards.
2. Tools that lack guards, equipment that poses a hazard, or defects in objects.
3. Poor housekeeping practices or conditions.
4. The presence of flammables, fire, and explosion hazards.
5. Walking and working surfaces that are hazardous.
6. Environmental hazards such as chemicals, radiation, and noise.
7. Placement of objects that protrude into aisles.
8. Hazards created by machines or objects that do not warn of movement.
9. Poor or inadequate lighting.
10. Unsafe personal protective equipment or clothing.
11. Inadequate or defective warning systems.
12. Plant layout hazards that provide close clearances or congestion.
13. Hazardous placement of storage products or items.
14. Holes, pits, shafts and elevated work areas.
15. Poor or inadequate maintenance procedures.

This list represents the most common unsafe workplace conditions, but it should not be considered a comprehensive list of all unsafe conditions that could be present in the workplace. When a JHA is being completed at the work station or work area, supervisors should focus on the actions of the worker as well as the physical conditions in which they work. Supervisors should specifically look for the presence of any item on the aforementioned list. For workers that assist in conducting safety inspections of the work area, the knowledge of what conditions to look for can be very helpful in injury prevention.

The remainder of this section discusses in detail each of the 15 types of unsafe conditions listed above, and provides examples of common injuries or incidents associated with each.

Lack of or Inadequate Safeguards

Unsafe machines are very common throughout industry. When injuries occur as a result of missing or inadequate guarding, the results are usually very serious. Even though machine-related injuries do not occur with the same frequency that other types of injuries occur, both the severity of the injury and its related costs are higher.

Bench grinders and power presses are the most common machines to be cited by OSHA for violations of their machine guarding standard. Management may say that a machine does not have proper safeguards because the machine was purchased without the proper guarding, or that maintenance departments or contractors removed the guards while working on the machine. Even though OSHA has been present for almost 30 years, many manufacturers of machines as well as employers fail to consider complete machine guarding for the product. Once removed by employees, guards are sometimes discarded and the machine remains unguarded for years.

If a worker is accustomed to operating machines without guards, it may become second nature to ignore the risk altogether. In many cases, the worker is not trained to identify unsafe conditions, but is trained to operate the machine. As a result, the worker lacks the training necessary to keep safety in the forefront of his attention, and the potential for an injury is more likely.

In addition to proper training, facilities must establish policies that require a machine lacking a guard to be immediately shut down and not operated until proper safeguards are provided.

Example of a Machine Injury:

A worker in an automotive parts manufacturing plant arrived at her machine at 7 a.m. to flare the ends of small sections of light gauge tubing. The night shift had performed maintenance on the machine and did not reinstall the guard. Despite a requirement to install all guards before starting the machine, the worker began to run parts. At 7:20 she amputated the first digit of her right hand. She had been holding the tubing in her right hand and had cycled the machine with a foot pedal.

Tools That Lack Guards, Equipment That Poses a Hazard, or Defects in Objects

Tools can lack safeguards and pose a hazard to the user. Circular saws are potentially dangerous and require skill on the part of the operator. The blade cover guard can be missing from the saw or tied up out of the way. Nail guns also contain guards that may be circumvented or defective. Other tools that are originally equipped with guarding may become unsafe over a period of time through wear, neglect, or damage.

Common tools can pose unexpected risks. Tips of screwdrivers can be distorted or broken. Hammers can have loose or splintered handles. Saws can have dull blades. Socket sets can be made of inferior products and shatter during use.

Tools and equipment can also have defects that can easily contribute to injuries. Consider the need for inspecting and correcting defects in: ladders, benches, stools, dollies, carts, grinders, and containers. A piece of equipment that is very common and potentially dangerous is the paper cutter. The situation depicted in Figure 5.1 can result in injury if the blade is not used properly. The photo depicts a situation in which the blade presents a hazard.

Tools should be evaluated when completing a JHA, and they must be on the routine inspection list. Look for the following defects or conditions when evaluating tools and objects: decay, wear, cracks, missing parts, fraying, poor design or construction, slipperiness, contains splinters, or is sharp-edged. All tools should be inspected for safety before using them in the workplace.

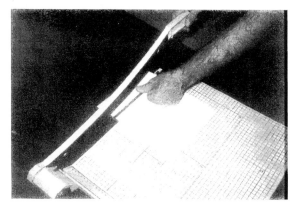

Figure 5.1 Hazards of a paper cutter
The situation in this photo has been simulated to demonstrate the potential hazards of a paper cutter.

Example of a Power Tool Injury:

A 7-inch disc on a portable hand grinder injured a worker in a steel fabricating plant. The guard for the disc was removed several weeks before and was not reinstalled. As the worker was grinding a steel plate he brought the grinder toward his right leg and the rotating disc cut into his leg. He received 15 stitches from the wound.

Poor Housekeeping Practice or Conditions

Proper housekeeping practices are essential in every workplace. Correct housekeeping improves efficiency, reduces injuries, and provides for a better-looking workplace. Poor housekeeping can easily deteriorate into traps that injure workers and visitors at the site, restrict productivity and present an image of uncaring management. Unsafe housekeeping can easily result in trips and falls, contusions, cuts, and abrasions. Workers can easily trip on materials left in walkways, on stairs, or at their workstations. Grease or oil on the floor can cause a fall that could easily result in a serious injury.

There should be assigned places for tools, brooms, mops, ladders, chains, hooks, cables, and cleaning supplies. There should also be a designated location for pieces of scrap, leftovers, and materials that have to be salvaged. Management should commit to a program of clean floors and a program that stresses that everything must be in its designated place when not in use. In the process of conducting a JHA, look at the housekeeping in the worker's area and correct deficiencies at that time.

Example of an Injury Associated With Poor Housekeeping Practices:

A warehouseman was unpacking boxes of automotive parts that were held together with plastic strapping. The first few straps required cutting with a knife to help open the box.

Once cut, the remaining bands could easily be slipped off of the boxes. The bands that were not cut were all over the floor by the worker. He failed to pick up the discarded bands and got both shoes caught in a band while walking away from the work area. This caused him to fall forward and strike his chin on the floor. He was carrying a few boxes at the time, and was not able to maintain his balance.

Flammables, Fire, and Explosion Hazards

The past 25 years have seen an increase in the amount of chemicals used in the workplace. Many of these chemicals are dangerous if not handled or stored properly. Chemicals can have low flashpoints, or be toxic or environmentally harmful. There is no question that fire and explosion can result from unsafe conditions and procedures.

There are safeguards that can be taken to protect workers and property from flammables and chemicals.

- When purchasing chemicals, always evaluate the chemical by looking at the material safety data sheet (MSDS) and choose products with higher flashpoints and exposure limits that comply with OSHA regulations.
- Ensure that the MSDSs are readily available for supervisors and workers. Everyone must be trained to recognize the hazards associated with the chemicals they are working with. Follow the guidelines of the product manufacturer and OSHA's Hazard Communication Standard.
- Ensure that safety devices are readily available for use on drums and containers containing flammables. Safety devices can prevent spills as well as fire and explosion. Check with equipment suppliers that handle safety bungs, spigots, grounding and bonding devices, and equip containers with these devices.
- Isolate and separate chemicals that are not compatible. Check with a chemical engineer or the manufacturer for tips on safe storage.
- Equip storage areas with the required sprinklers and fire safety features as required by the National Fire Protection Association (NFPA), Factory Mutual Engineering, (FM), or your local fire marshal.
- Train workers to prevent fires from occurring, and to fight fires when they do occur. Organize fire teams to search out potential fire hazards and correct these hazards. Use only trained professionals to fight fires.
- Ensure that the proper personal protective equipment is readily available in the event of a chemical spill. Have spill cleanup kits and procedures available. A JHA that should be on every list is "how to safely clean up a chemical spill."

Example of a Fire Safety Incident:

A forklift operator was driving a propane-powered vehicle that ran out of fuel. He was outside at the time and removed the empty tank and walked to the tank storage area. He was smoking at the time and failed to recognize the hazard. A tank in the storage area was leaking propane gas unbeknownst to the operator. His cigarette somehow started a fire in the storage tanks. The local fire department was summoned to extinguish this large blaze.

Hazardous Walking and Working Surfaces

Slips and falls are one of the biggest producers of injuries in the workplace and in the home. Employees can stumble on sections of the floor that are off-set, on cracks in the floor, or at curbs that are not identified properly. Employees can fall on product, scrap, tools, hoses, wires, off of ledges, from scaffolds, into shafts or holes, and on grease or oil.

Stairs must be included in the list of potentially hazardous areas. There could be product on the stairs, lighting may be poor, and / or the hazard may be unseen. Hand railings may be missing, defective, or inadequate. The pitch of the stairs may not meet standards.

Consult OSHA regulations to determine and create safe walking and working surfaces. When new facilities are being developed or existing operations are being modified, the focus should be on compliance with the OSHA requirements. Where possible, repair holes, cracks, and floor off-sets to prevent a fall. If the problems with the floor cannot be corrected, paint the hazard to identify it or place a barricade to prevent an incident.

An Example of a Slip and Fall Hazard:

Two workers were moving furniture from a building onto a truck. They were facing each other while carrying a heavy couch and did not notice the edge of the curb. The concrete was old and the edge of the curb was crumbling. The hazard had not been repaired for several years and was not highlighted with cones or paint. The worker on the lead end of the couch stumbled on the curb and fell backward with the couch falling on top of him. He fractured his ankle and broke two ribs in the fall.

Environmental Hazards Such as Chemicals, Radiation, and Noise

Fumes, vapors, mists, solvents, gases, radiation, and noise in the workplace can harm employees. Situations where there can be a lack of oxygen, such as a confined space, must also be considered. Extreme heat or cold can also create unsafe working conditions. Many of these hazards are unseen and the worker may not realize that these unsafe conditions even exist. Radiation is one of the more serious unseen hazards and requires stringent regulation.

When JHA's are being conducted, management must consider the potential workplace hazards that can harm the lungs, skin, or other body organs. Of the various chemicals and workplace exposures that can cause harm to workers, some 90 percent are inhaled into the body. Eight percent are absorbed into the skin, and 2 percent are ingested. Once a chemical goes into the body, it can easily enter the bloodstream and cause harm to target organs such as the kidneys or liver. Dusts and fibers can easily make their way to the lungs if the size of the contaminant is microscopic. Workers can easily be harmed by any of these environmental hazards.

Environmental conditions must be evaluated regularly to monitor exposure. It is the responsibility of management to identify the chemicals people will be working with, measure the air so harmful levels are not exceeded, and take steps to reduce harmful levels. The installation of ventilation, the use of personal protective equipment, the substitution of less hazardous chemicals, and employee training can all be combined to form a safer workplace.

Example of an Environmental Hazard:

A forklift operator complained of a headache to his supervisor during the early part of his shift. Later that afternoon he became nauseous and had to go to the emergency room. The clinic diagnosed the problem as carbon monoxide poisoning. Internal combustion engines produce carbon monoxide. Engines must be checked and properly tuned to keep exhaust emissions to their appropriate levels. Catalytic converters can be added to forklifts to reduce emissions, or, electrically operated forklifts can be substituted for those with internal combustion engines.

Placement of Objects That Protrude into Aisles and Working Areas

People don't deliberately place objects or obstructions in aisles or by machines to harm others, yet some would say that it sometimes seems that way. Injury can occur because the person is not able to see the hazard or because the hazard is in a dark area. A valve at ankle level that protrudes into an aisle, a piece of lumber that contains a rusty nail on its end or a sharp piece of metal protruding from a scrap bin can easily cause an injury.

As with other hazards, management must make an effort to inspect for these hazards and correct them when discovered. In some cases, the maintenance department must be used to search out and correct fixed hazards that are potentially harmful. Evidence of torn clothing or mild abrasions often indicates the presence of a hazard that can result in serious injury.

Example of a "Protrusion into an Aisle" Injury:

A plant manager was escorting visitors through a factory and walked directly into a piece of metal banding on a steel coil. The banding was at face level and as the plant manager rounded a corner was busy talking to the visitors. He walked into the sharp edge of the metal band and suffered a severe cut on his left cheek.

Figure 5.2 identifies a metal barrier guard that was designed to protect workers from lift trucks. One of the problems with the barrier was that the ends protruded into a walking aisle. To correct this, management padded the ends of the barrier.

Hazards Created By Machines or Objects That Give No Warning of Movement

Machines, tools, cylinders, vehicles, and product can slide, drift, fall, bend, roll or start up unexpectedly. Workers injured in situations such as these usually report that they had no idea that the load or piece of material was going to move. If a worker is not aware that something is going to move, they can't protect themselves from injury.

Injury prevention usually stresses the known hazards, but unexpected movement is usually an unknown factor. Supervisors and safety commit-

Figure 5.2 Protective padding on a barrier
To prevent workers from striking their knees on the barrier guard, rubber insulation was placed on the edges.

tee representatives can best protect against these hazards by assuring that mobile equipment in the workplace have proper warning devices, identifying objects that may become mobile as a result of design or error, and training employees in proper use of equipment.

Examples of Unexpected Movement Injuries:

A stock chaser that is backing up can injure a worker if the vehicle has no back up alarm. A machine can start up unexpectedly and catch a worker's hand between two rollers. A vehicle can be parked on an incline and not have the wheels blocked, or it could even be in gear. Maintenance workers or setup crews can be injured at an operation because the power source is not locked out. An oxygen cylinder can topple over on a technician because the cylinder is not secured. Materials can be stored or positioned in such a way that allows them to move unexpectedly.

Poor or Inadequate Lighting

Poor lighting can be an easily avoidable cause of injury. Employees can fall down stairs, step in holes, and select the wrong chemical because of inadequate lighting. The cause of insufficient lighting can range from burned-out bulbs, defective ballasts, inadequate bulb size, and the placement of the light fixture.

In warehouse operations, having the lighting in the middle of the aisle is ideal. However, when storage racks are relocated, the lighting fixtures are not always relocated. As a result, workers are partially in the dark while performing their jobs in the aisles.

Proper lighting has a bearing on safety, productivity, and quality. If the correct parts are to be selected for shipment, or precise measurements are to be made when machining a part, correct lighting is important. In addition, it is hard to maintain high standards of housekeeping in darker areas.

At the least, inspect lighting to assure that it meets OSHA requirements. When planning programs to improve lighting, do not forget emergency lighting. In most cases, the fire marshal requires emergency lights. Include these lights on the safety inspection list.

Example of an Injury-Related to Inadequate Lighting:

Two workers were assisting a mobile crane operator to move sheets of steel from the side of a building to the inside. The work was being performed in the evening hours on the second shift. One of the workers was waiting for the crane to boom over the load with the chains and clamps that were needed to lift the load. Because of the darkness, the worker did not see the swinging chains and clamps in time. He was struck in the jaw by the hardware and lost several teeth. This job was usually done during the daylight hours, but the sheets of steel were needed for the next shift. All the workers had to light their work area was the corner street light.

Unsafe Personal Protective Equipment or Clothing

Some refer to this unsafe condition as "unsafe personal attire." Unsafe clothing can consist of shoes with unstitched soles, shoes with the steel toe about to fall out, oil soaked pants or shirts, worn heels on shoes, loose clothing, rags tied around the waist being used as an apron, rings, earrings, torn gloves, hard hats with the suspension modified, and neckties. Unsafe attire can get the employee in trouble if part of his or her clothing is caught on a machine, or snags on a fixed object while walking.

Personal protective equipment and clothing should be inspected for safety as strenuously as the machines the employee is operating. When observing a worker while performing a JHA, look at his clothing. Do you see any clothing problems that were mentioned above? Some workers are not as quick to repair or replace defective or inadequate equipment. In most cases, management must insist on it to keep the worker safe.

Example of an Injury Caused by Unsafe Attire:

A worker was walking through a machine shop during lunch when the large rag tucked into his belt caught on a ladle handle. A coworker was involved in a personal job of melting lead for fishing sinkers and the large rag pulled the molten ladle over onto the foot of the passing worker. He suffered major burns to his foot.

Inadequate or Defective Warning Systems

Employees can be injured by machines or processes that lack a means of warning. Overhead cranes, backing forklifts or earth-moving machines and conveyors are usually equipped with devices that will sound an alarm when they are moving. In some cases, the warning device is missing or in need of repair.

Warning devices include lights, buzzers, and other audio or visual signals. A flashing light will be more conspicuous than a stationary one. An audible alarm and a flashing light are more noticeable than the flashing light alone. Additional methods that can be used to provide warnings are guards or spotters. Be sure to brief these individuals so they know what their assignment is. Flags, barricades, cones, directional arrows, traffic lights, and painted lines can also successfully warn employees of hazards.

Provide a warning system that matches the likelihood and severity of injuries that may occur. Inspect warning devices on a regular basis. When performing a JHA, be sure to check that the alarm or light is functional.

Example of an Injury Caused by Inadequate Warning Systems:

A forklift operator was changing a propane tank on his machine when he was suddenly pinned to the counterweight by another backing forklift. The backing forklift was not equipped with a back up alarm. The impact of the collision resulted in the death of the operator on foot.

Plant Layout That Contributes to Close Clearances and Congestion

When little consideration is given to the working area around the employee, the closeness of the operation can be the cause of injury. Much of this can be prevented by pre planning the layout of machines and pedestrian walkways. Dock areas where trucks, forklifts and other powered equipment are constantly moving product in and out of the building usually need good planning to prevent injuries. The National Safety Council estimates that between 10 and 25 percent of all injuries occur at the dock. Adding safeguards can eliminate many hazards.

Truck drivers can be protected from moving powered industrial trucks by having them report to an area at the dock that is protected by posts and rails. Yellow walkway lines and the appropriate signs can also help guide them to offices that they must visit. Convex mirrors can provide warning of approaching vehicles. Each year visitors to the dock area of a plant are injured because of the busy activity, congestion, and unfamiliarity of the location.

To protect truck drivers at a pick up area of a dock by prohibiting forklift traffic in the area, management mounted a sign reading: "Forklifts NOT Permitted in Truck Loading Area."

Where workers have to squeeze around machines, obstacles, or production in process, the tight areas and potential of being squeezed by a machine or process must be considered. In most cases, remove and relocate the obstacle or barricade the area to prevent passage by anyone. Cross sections of a factory could be endangering workers because of heavy vehicle traffic and pedestrian activity. Traffic signals may be needed along with signs, lighting, mirrors, and designated walkways.

Example of a Close Clearance Injury:

A worker used the same walkway each day to go on his break and return the same way. He would walk between a shaper and a steel beam rather than add a longer walk on a designated walkway. His shortcut proved to be a problem one morning when he did not count on the bed of the shaper to be moving. He was pinned between the back of the shaper and the steel beam. He suffered serious internal injuries.

Hazardous Placement of Storage, Products, or Other Items

Because of a lack of floor space and poor planning, worksites can sometimes create production hazards and safety issues that involve materials storage. Boxes piled too high can easily collapse or be struck by an overhead crane or lift truck. Materials can block exits, cause congestion, and block the vision of those on foot. Emergency equipment, doors, restrooms, directional signs, and sprinkler heads can all be rendered ineffective by unsafe storage.

Additional concerns for this type of unsafe condition involves the placement of scaffolds, ladders, and workstations in areas where they could be a problem for everyone. Workers should not have to navigate walkways and other parts of the plant by having to bend over and walk under obstacles. In other cases, materials are in walkways and everyone using the walkway must constantly step over obstacles. Also, doors that only partially open are a problem when they are blocked.

Pre planning of the product flow can help to relieve congestion. There comes a point in time however when a plant must expand to accommodate the product movement within the building.

Example of an Injury Caused by the Hazardous Placement of a Ladder:

An emergency light was not working and a supervisor asked a maintenance person to replace the light. The light was mounted about 10 feet high over the main door to the locker room. No one took time to place a spotter, cones, or a barricade in place while the work was being completed. Workers were going in and out of the locker room while the project was taking place. While attempting to mount the new emergency light fixture, the maintenance worker dropped it. The falling fixture struck a worker on the shoulder causing a serious injury.

Holes, Pits, Shafts and Other Elevated Work Areas

Most times when someone falls into a hole or pit, the results are serious. Undoubtedly, these are hazards that must be taken seriously. Some holes or pits are permanent items at the worksite and many employees recognize their presence. Visitors and new employees do not have the same level of awareness that the plant workers have. In some cases, the visitor may be in an industrial setting for the first time.

These types of hazards require barricading as well as alerts and warnings of their presence. If at all possible, remove the hazard completely.

Example of an Injury Caused by a Pit:

An auto mechanic was leading a customer to her vehicle to show her the defective parts. They had to walk by an open pit that was once used to service vehicles. The perimeter of the pit was not guarded in any way. The customer was preoccupied with the thought of her vehicle and did not notice the raised kick plate around the pit. She stumbled on the kick plate and fell into the pit where she suffered multiple injuries.

Poor or Inadequate Maintenance Procedures

To keep a worksite or factory running, on-going maintenance is essential. Unsafe conditions can be created when work processes are not completed correctly. If guards are not replaced or installed properly, drain plugs are not tightened, tools are not removed after the job has been completed, or remnants of grease or oil are left on the floor, serious injury could result.

As a means of personal protection, maintenance staff must always use a zero energy program when working on machines. The proper handling of confined space entry and protection against harmful chemicals should also be a part of every program in a maintenance department. The use of schedules to provide preventive maintenance to machines can keep an injury from occurring because the exposures during unnecessary repair and replacement would not be present. Management has the ultimate responsibility of employing mechanics that are qualified and ensuring that maintenance procedures are correct.

Example of an Injury Caused by Improper Maintenance Procedures:

A mechanic replaced a thermostat in a radiator hose and installed it backwards. The car overheated and the mechanic removed the radiator cap without allowing the car to cool down. He was burned by the erupting anti-freeze from the radiator.

Unsafe Behavior

Unsafe behavior is any action by one or more employees that increases the chances of an injury or illness. There have been studies that expose the theory that a large number of injuries are the result of unsafe behavior.

When injuries occur, do not always assume that the worker is solely at fault—even if he brought about the injury as a result of something he did or failed to do. Supervisors must ask if the worker was properly trained. Determining the cause of the injury does not end by identifying what was done or not done. Management must also determine why it was done or not done. It is necessary to search for the causes of injuries by investigating the physical action, lack of action, and the total environment the worker is exposed to. If the investigation is conducted properly, in almost every case the causes will be more apparent when both behavior and work environment are evaluated.

Of course, every safety director and plant manager knows that there are a few individuals in the workplace that have developed bad habits along the way, despite any training they may have received. However, according to author Sydney J. Harris, "If you start to learn something the wrong way (which is usually the easiest way), the longer you practice, the more ingrained become your bad habits, and the longer it takes to correct them and get on the proper path." Good habits and safe behavior should be emphasized in initial training of employees because changing habits can be difficult.

Do not expect all workers to readily accept any rule or procedure that is given to them. In many cases, the need for a specific rule has to be sold to workers—especially if they will be required to deviate from their past procedures. It is wrong to say, "Here is a new rule. I expect it to be carried out because it is good for you." Performing a JHA allows workers the opportunity to develop the new rules and procedures they will then be expected to follow. Through the JHA, the worker will be aware of the hazards in his workplace and can suggest his own alternative procedures. It is important that the employee is aware that his input is valuable and that whatever procedure or rule is adopted for the job is done so in his best interest. In this way, the JHA system provides an excellent means for teaching and communicating.

Consider the safety and legal reasons that exist for wearing seatbelts in automobiles. It is logical to accept the fact that seat belts in cars save lives and help to prevent injury. There is a law of physics that states that a body in motion will continue in motion unless stopped by another force. When a vehicle comes to a sudden or unexpected stop, passengers not wearing seatbelts will continue to travel forward at the same speed that the car was traveling before it came to a halt. In addition, not wearing a seatbelt carries a financial penalty if the driver is given a ticket. There are both health-related and financial reasons for wearing seatbelts, but many passengers still refuse to wear them.

Similarly, all employees may not follow rules or use procedures that were established for obvious health and regulatory reasons. Risk-taking sometimes becomes the norm because protective equipment (including seatbelts) that must be worn is uncomfortable or the prescribed procedure takes longer. In addition, some employees, like some drivers, may not follow prescribed rules or procedures because they are not convinced that an injury will follow or they believe that the likelihood of injury is so small given the situation or their experience and skills that they are willing to risk it.

Unsafe behavior is not limited to workers. Many members of management take unnecessary risks and risk injury to themselves and employees. When supervisors do not repair equipment or ignore established safety rules and procedures, then management demonstrates the same unsafe behavior. Many times this demonstration of breaking the rules or acting foolishly is performed in front of workers. If management was ever trying to send a message on how they feel about safety, the use of unsafe behavior can easily convince everyone that safety is not important to them or that the observed activity is an acceptable and safe act. On the other hand, when management believes in safety and holds the welfare of all the workers as a company value, it is more likely that employees will respond the same way.

Supervisors, plant managers, safety committee representatives, and workers should be aware of common unsafe behavior, including the following:

1. Failure to wear prescribed personal protective equipment.
2. Unauthorized use of equipment or tools.
3. Improper lifting, carrying, loading, or sorting.
4. Failure to use lockout/tagout procedures when working on equipment or devices.
5. Use of defective tools or parts.
6. Disabling or removing guards or electronic devices.
7. Failure to warn or signal of movement.
8. Failure to abide by speed limits or load limits.
9. Distracting others that are working.
10. Working on moving dangerous equipment.

This list represents the most common unsafe workplace behavior, but it should not be considered to be a complete list of all unsafe behaviors that could be present in the workplace. When a JHA is being completed at the workstation or work area, supervisors should carefully observe the worker's behavior, and should specifically look for the presence of any item on this list. They should also use good judgement in recognizing similar forms of unsafe behavior that do not appear on the list.

The remainder of this section discusses each of the 10 types of unsafe behaviors listed above in detail, and provides examples of common injuries or incidents associated with each.

Failure to Wear Prescribed Personal Protective Equipment

Figure 5.3 Proper PPE not worn Gloves should be worn whenever pulling on a sharp object

When workplace hazards cannot be guarded or eliminated, sometimes the next best thing to do is to require the use of personal protective equipment (PPE). OSHA regulations require certain PPE for specific jobs, and management should choose additional equipment based on the nature of the hazard. There are numerous hazards that can cause serious injury that only

PPE can prevent. Wearing PPE improperly, wearing the incorrect PPE, and not wearing the required PPE at the correct time can injure workers.

The difficulty is having all of the workers and members of management wear the PPE where and when required. PPE must be chosen for reliability, protection, and comfort. It is not unusual, especially during warm weather, for workers to disregard PPE in order to increase their personal comfort.

When conducting a JHA, a part of the process requires that the correct PPE be identified for the job. When required PPE is not worn, immediate action on the part of management is needed. Use the correct protection for the specific hazard.

Example of an Injury Caused by Not Wearing Proper Personal Protective Equipment:

An auto mechanic was using a cutting torch to remove clamps on a defective muffler. He was not wearing the prescribed cutting goggles. His rationale was that the job was going to take only a few moments. Despite the squinting of his eyes to shield them from the sparks and molten metal, a blob of hot metal landed on his cornea and he had to have emergency care for the injured eye. He had to miss a month of work as a result of the injury.

Unauthorized Use of Equipment or Tools

Workers sometimes use equipment that they have not been trained for. Employees have been known to operate powered industrial trucks, overhead cranes, mobile cranes, and power tools without authorization and / or training. In most cases, the worker has not been trained to operate the equipment or the worker is not qualified for that particular job description.

One of the most frequent examples of unauthorized use of equipment is untrained workers operating powered equipment—sometimes with tragic results. Additional risks are present when workers throw switches, turn valves, or provide signals without authority.

During new employee orientation it is essential to alert each new hire that there are restrictions on machine operations and that the unauthorized operation of equipment could possibly result in discipline or discharge. When conducting the JHA, supervisors should confirm that the worker has received adequate training for any equipment he is observed using.

Example of an Injury That Resulted From Unauthorized Use of Equipment:

A warehouse employee took it upon himself to move a pallet load of product with an electric pallet truck. He was not authorized to operate any powered equipment and was told of this restriction during his orientation. While moving the pallet he struck another employee in the Achilles tendon, seriously injuring him.

Improper Lifting, Carrying, Loading or Sorting

With the recent national emphasis on workplace ergonomics, preventing injuries from manual lifting and handling of product and loads is essential. Research and numerous studies offer evidence regarding the correct ways to carry, lift, load, and manually handle product.

Management must train workers in the correct techniques to prevent injury and seek out potential ergonomic hazards.

When workers are observed performing jobs in an ergonomically unsafe manner, corrective action must be taken. To reduce injury, management should consider the purchase of powered equipment to remove the need to manually handle product. However, despite the increased use of powered equipment to move and handle materials, there will always be some need to manually handle product. When evaluating work methods, the person conducting the JHA should look for correct postures and provide corrective solutions if necessary.

Example of an Injury Caused by Improper Lifting:

Two maintenance workers were attempting to carry and set down a large metal plate. As they were attempting to set it down their signals became crossed and instead of lowering the load together, one worker dropped his end and the other worker suffered a disabling back injury because he was still holding on to his end of the plate.

Failure to Use Lockout / Tagout Procedures When Working on Equipment or Devices

When working on any equipment that has electrical, hydraulic, pneumatic, steam, or stored energy, danger exists if the worker does not take proper steps to protect himself. Sometimes something as simple as a 2 dollar lock can save a worker's life. Each year, there are approximately 140 fatalities as a direct result of not using lockout / tagout procedures.

Everyone must be trained to respect the hazards associated with various forms of energy. Each JHA that is completed on tasks that involve various forms of energy must focus on the exact procedures to be followed when working with these forms of energy. Where possible, relocate oil feeds or grease fittings so the running machine can be lubricated in a manner that does not endanger the worker. For comprehensive information and guidelines on attaining a zero energy state, look at OSHA's 29 CFR 1910.147 for the control of hazardous energy.

Example of an Injury From a Machine for Failing to Eliminate the Power Source

A maintenance worker was asked to check a leak in a propane cylinder hose on a forklift. He failed to use the valve on the cylinder to stop the pressure from coming out of the hose. He was not wearing PPE and suffered frost burns to his hands when the gas released while he was removing the hose with his hand.

Unsafe Use of Tools and Parts

Workers are frequently observed using tools for tasks for which they were not designed or intended. Some common examples are: using a screwdriver as a chisel and striking it with a hammer; using a portable circular saw while tying the blade guard up and out of the way; and using a shovel as a hammer. Tools are designed to be used a certain way for the task at hand. When workers deviate from the correct procedure, they risk injury.

The use of improper tools should be corrected during the JHA. However, many employees know the correct use of the tool, but will risk injury from improper use rather than taking the time to get the right equipment. This behavior is usually discovered through informal observations in the workplace.

Example of an Injury Caused by the Misuse of a Tool:

An auto mechanic was removing a frozen bearing from an axle and was using a hammer and chisel to loosen the bearing. The chisel was seriously spalled, and while striking the chisel a steel fragment came off and struck the mechanic on the cheek. The sliver of steel imbedded itself in his skin. The supervisor investigating the injury reminded the mechanic that he was fortunate he wasn't struck in the eye by the fragment since he was not wearing safety glasses.

The employee in Figure 5.4 is in danger of cutting his thumb because he is not using the cutting knife properly.

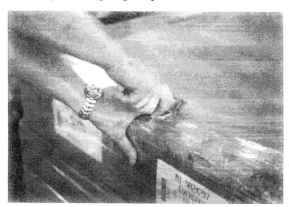

Figure 5.4 Cutting a box in an unsafe manner
This worker is sure to cut his thumb in this simulated photo.

Disabling or Removing Guards and Electronic Devices

Guards on machines are there to protect the worker—not the machine. Guards are there because a serious hazard exists. However, operators often deliberately remove guards because the operator believes they are in the way or are slowing down production. Guards can also be removed by maintenance departments or setup workers that operate the machine. When maintenance and setup workers remove the guards as a result of servicing the machine, the intent is to replace them as soon as the work is complete. In some cases, the guards are not placed back on the machine, or are placed on the machine incorrectly. In each case, the operator is exposed to potential harm.

Safeguards that can be removed or tampered with include: ground prongs on electrical plugs, bonding and grounding wires on flammables, and tags that read "danger" or "warning." Some workers have been known to go so far as to use bolt cutters on locks so they could run a few more parts on a defective machine. In other cases, operators tie back guards or place tape or wire on limit switches or interlocks, thus rendering the guards or electronic protection worthless. Thus, caution must be exercised when piecework and incentive programs are used for production quotas on machines.

The removal of guards is not behavior that is typically observed during a JHA, but their absence should be noted during the observation and corrective actions taken. Again, it is necessary to determine not just what the hazard is, but why the hazard exists. Determine if the worker, maintenance, or setup workers removed the guard, and provide proper training if necessary.

Example of an Injury Caused by Removing a Guard:

A worker in a steel fabricating plant came in to run his machine during the second shift. He was unaware that the worker on the day shift had placed a small wire on an interlock and disengaged it. During the shift, the operator had to free a jammed part from the dies of the machine. After releasing and removing the part, the machine cycled unexpectedly and crushed the right hand of the operator.

Failure to Signal or Warn of Movement

Machines or equipment can move unexpectedly at any time during work. If there is an operator at or on the machine, it is their duty to warn others in advance so they can stand clear. Just as equipment that may became mobile requires warning systems, operators of mobile equipment must also warn workers in the vicinity of their intentions.

The operators that control overhead cab cranes rely on a "hook-up man" on the factory floor for directions. It is a standard procedure for the crane operator to move loads, chains, or clamps only when directed to do so. In fact, the operator can be operating at a height or angle that would not afford him the opportunity to always see if a person is clear of a load before moving. Workers have been injured when operators took it upon themselves to move the crane without being signaled to do so.

Forklift operators are also responsible for using a horn or verbal alert before moving. OSHA does not require back-up alarms on lift trucks, and it is necessary to warn anyone near a truck that it is moving. It is wrong to assume that the operator of the vehicle has the right-of-way. In fact, before moving, the operator should make eye contact with the person to show his intentions. Alarms and verbal warnings are important. At machines that are being serviced, everyone should be alerted to stand clear before the machine is activated, and power should not be applied until everyone is clear of the machine.

Take advantages of horns, sirens, lights, hand signals, and verbal instructions to keep others safe. Be sure that operators and workers in an area are aware of and heed the proper warnings.

Examples of Incidents Where the Employee Failed to Warn:

A crane hook-up man was loading flat bars on a rack near a machine and was also involved in a discussion with a co-worker. After the load was set on the rack, a signal was given to the crane operator to lift the chains. Instead of paying attention to the chains and the hooks on their ends, he was engaged in conversation. The crane operator could not see the ends of the chains because of the rack. The chain hooks caught on more flat bars that were stored on the same rack. The machine operator sensed what was about to happen and pulled the hook-up man out of the way, narrowly missing being struck by hundreds of pounds of steel bars. He should not have signaled the crane operator to raise the chains until he was sure that the chain hooks were clear.

A worker fractured his wrist when he had his hand jammed against and was caught between a metal post and the wooden post on a cart that he was holding. A forklift operator backed up without warning and struck his cart. This injury is simulated in Figure 5.5.

Figure 5.5 Trapping a wrist between a cart handle and a post

A hand or wrist is vulnerable to being trapped between two objects.

Failure to Abide by Speed Limits and Load Limits

Driving too fast or overloading can be a problem in the workplace. Just as it is important to obey speed limits on the road for safety, similar rules in the workplace should be observed. Various pieces of power equipment can easily be driven too fast and other equipment overloaded if safety measures are not established.

Despite the fact that a forklift can usually operate at 5-10 miles per hour, the slightest bump or collision can injure workers or damage property. Trying to stop on wet pavement is another problem with forklifts. Operators must slow down where slick floors are present. In parking lots, it is important to drive cautiously to avoid ruts and other vehicles. Forklifts often take longer to stop than cars, so the speed must be low enough to allow for gradual braking and stopping.

Stationary machines that have movable parts also require the establishment of safe operating speeds. Conveyors can be tampered with to gain more speed. Operators may take shortcuts to increase speed on machines—especially if bonus or incentive plans are a consideration.

Overloading is another problem that can injure the worker or damage property. If a piece of machinery is rated for a certain capacity, the manufacturer knows the operating limits of the machine. Employers should identify capacities and ensure that the numbers are in full view of the operator. In other cases, the employee must use sound judgement to avoid overloading. As an example, before using a wheelbarrow, carrying lumber or shingles, or manually handling any other load, first size it up. If the load looks too heavy, get help or reduce the weight by removing some of the material. Apply this same rule to forklifts, cranes, conveyors, dollies, chains, wire rope, and carts.

When observing a job, take notice of the speed at which workers operate machines, and the amount of material loaded to determine that safety limits are not exceeded.

Example of Failure to Abide by Speed Limits:

A forklift operator was traveling from one building to another while driving through the employee parking lot. He was traveling too fast and did not see the large pothole in the roadway. His left front tire went into the hole and the forklift tipped over onto its side, throwing the operator from the seat and onto the pavement. The overhead guard pinned his legs, and he suffered multiple fractures to his legs. He was not wearing his seat belt.

Distracting Others That are Working

Workers are injured each year when others distract them in the workplace. Workers can startle, tease, or distract fellow workers while on the job. Some people love to play practical jokes on their neighbors or fellow workers. Some examples of distractions that can be found in the workplace include: throwing snowballs, throwing water or water balloons, firecrackers, putting grease on handles of machines and chasing or teasing a fellow worker.

Although it is clearly impossible and not even desirable for management to eliminate humor or communication between workers, supervisors should send a clear message on the dangers of certain behavior or "jokes."

Example of an Injury Caused by a Practical Joke:

A worker wanted to play a practical joke on a crane operator so he placed blobs of grease on the rungs of the access ladder. The operator did get grease on his hands when he started to climb the ladder. However, the practical joker did not realize that the operator would continue to climb the ladder. The crane operator's foot slipped off of a ladder rung and he injured his kneecap.

Working on Moving or Dangerous Equipment

Workers are frequently injured by failing to observe safety measures when working on moving equipment. Dangerous behavior includes failing to stand clear of moving cranes, powered carts, transfer buggies, and other powered equipment. Failure to control hazardous energy during this process is a problem. Hanging on to a chain while a crane is lifting the chain can have serious consequences.

Insist that workers allow the device to operate without being on the equipment and follow lockout procedures.

Example of an Injury Caused by a Worker Failing to Stand Clear:

A maintenance worker was called to check a noisy bearing on a gantry crane wheel. Instead of walking next to the crane wheel as it was being moved and using a long-handled oilcan, he rode on the frame of the crane and caught his hand between the wheel and structure. He fractured several fingers.

Checklists for Assessing Common Workplace Hazards

The following list of common workplace hazards should be assessed when completing a JHA. The list is by no means complete, but it provides a significant number of reminders that should be considered by the supervisor. The individual items are listed in the negative, meaning that the observer would detect the defective item so he or she could more easily recognize it. The provided photographs depict hazards that should be easily identified during a job observation.

Electrical Hazards

❑ Bare wires, poor connections, lack of ground fault protection, or a missing ground prong on a wall receptacle.

❑ Wiring not in compliance with the National Electric Code.

❑ Guarding/shielding not in place to protect anyone from live parts.

❑ Circuit breaker boxes not evaluated to ensure that all of the switches have been properly identified.

❑ A lack of hardware needed to perform safe lockout procedures.

❑ Electrical boxes not identified with the voltage and function.

❑ Extension cords lying across the walkway or in a walking/working area.

❑ Extension cords used as fixed wiring.

❑ Interlocks not functioning, are missing, or have been tampered with.

❑ Lack of a formal lockout / tagout program.

❑ Lack of training for individuals performing lockout / tagout.

❑ Incomplete or incorrect information on a lockout tag.

❑ Lockout tag missing from the lock or device.

❑ Outlets not tested for polarity or integrity.

❑ Lighting in areas where electrical work must be completed is inadequate.

❑ Areas in front of power panels do not have a minimum of three feet of clearance.

❑ Maintenance personnel not wearing the correct insulated shoes while performing electrical work.

❑ Electrical systems lack or have inadequate inspection procedures.

Machine Guarding

❑ Signs not at or on machines to warn of hazards.

❑ Guards missing, partially in place, or improper for the machine.

❑ Inspections of machines not being completed or inadequate.

❑ Personal protective equipment not being used or being used improperly.

❑ Housekeeping at the machine working area is in need of correcting.

❑ Electrical wiring on or at the machine is in need of correction.

❑ Employees not trained to perform machine setup as required.

❑ Interlocks defective, missing, or contain evidence of tampering.

Flammable Liquids

❑ Material safety data sheets are not available for flammables.

❑ Containers, which hold flammables, are not marked or identified properly.

❏ Grounding or bonding devices are missing from flammable containers.

❏ Safer chemicals not being used as replacements for more toxic or flammable materials.

❏ Areas in which flammables are being used are capable of producing ignition.

❏ Fire extinguishing equipment not readily available, the incorrect extinguishers are available or the employee has not been trained to fight fires.

❏ Flammable liquid cans are not equipped with flame arrestors and self-closing lids.

❏ Powered industrial trucks are not handling flammables safely.

❏ Spill clean up kits are not available in the event of an emergency situation.

❏ Ventilation is not working to safely evacuate chemicals / flammables.

Housekeeping

❏ The working area where the JHA is being prepared is in need of attention.

❏ There are patches of oil, water, ice, sand, grease, granular materials, or other fluids on the floor.

❏ Brooms, shovels and other cleaning tools are not available.

❏ Materials or product are in the working area and interfere with the job.

❏ The employee is not given enough time to properly clean up his working area.

❏ Lighting is insufficient to illuminate the job and the lights are not working properly.

❏ Machines need to be wiped off and are unsightly.

❏ Management must assign cleaning responsibilities to workers in the department.

❏ Floors have inadequate drainage and are not equipped with non-skid surfaces.

In Figure 5.6, a broken pallet board with a protruding nail can easily cause an injury.

Overhead Cranes and Hoists

❏ Wire ropes are not a part of a formal inspection program.

❏ Limit switches are not being inspected or tested as required.

❏ Loads being lifted exceed the capacity of the crane.

❏ Workers are not positioning the crane block directly above the load and are performing side pulls.

❏ The pendant box control buttons are not marked to identify their function.

Figure 5.6 Trip hazard
Broken pallet boards can be a trip hazard as well as having the potential for nails. Broken pallets must be replaced.

❑ The pendant box is not equipped with a supporting chain or cable to prevent the control wires from being pulled out of the box.

❑ The crane or hoist capacity is not marked on the sides of the bridge or hoist.

❑ The capacity of each crane block is not marked on the sides of the blocks.

❑ The crane or hoist-lifting hook has a missing or broken latch.

❑ The annual inspection of the crane or hoist has not taken place.

❑ The alarm that sounds when the crane is moving is not functioning.

❑ Workers have not been properly instructed on how to safely lift and handle loads.

❑ Clamps and spreader bars are not being inspected on a regular basis.

Fire Fighting Equipment

❑ Fire extinguishers are blocked and not accessible for fire fighting.

❑ Fire extinguishers lack one or more of the following: an inspection tag, a seal on the pin, a pin on the handle, a full charge, and / or an inspection.

❑ Employees are not trained in fire fighting.

❑ Signs identifying fire extinguisher locations are missing or improper.

❑ Fire extinguishers are not being inspected annually.

❑ Fire hoses are not being inspected and tested.

❑ Fire hoses are blocked and are not accessible.

❑ Sprinkler systems are not being tested on a regular basis.

❑ The storage of materials is too close to the sprinkler heads.

❑ A supply of extra sprinkler heads with the correct temperature setting is not available.

Oxygen / Acetylene Use and Storage

❑ Compressed gas cylinders are not capped, secured and separated properly.

❑ No smoking signs are not placed at or near the cylinder storage area.

❑ Oxy/Acetylene torches are not being used correctly.

❑ Workers have not been trained in the safe use of oxy/Acetylene systems.

❑ Cutting goggles are not available or are in need of repair or replacement.

❑ There is no fire extinguisher on the portable cylinder cart as required by state law.

❑ Cylinder gauges are set for the incorrect pressure or have a broken glass window.

❑ Reverse-flow check valves are not installed on the cutting torch.

❑ The oxy/acetylene hoses show signs of cracking/aging and need to be replaced.

Personal Protective Equipment (PPE)

❏ The worker being observed is not wearing the prescribed safety gear such as; safety glasses, hard hat, steel toe shoes, hearing protection, gloves and respirator.

❏ PPE is in need of repair or replacement.

❏ PPE is being worn improperly.

❏ The incorrect PPE is being worn.

Conveyors

❏ Pinch points are present on the conveyor(s) for the job being studied.

❏ The conveyor(s) are not set at the correct working height for the worker.

❏ Warning signs intended to alert the worker of conveyor hazards are not available or are in need of replacement.

❏ There is no means of access to go over the top of the conveyor, (stairs and railing).

❏ The emergency shut-off switch / cord is not available or is not working properly.

❏ For conveyors that roll, the wheels are broken or not functioning properly.

❏ Workers are manually handling sections of portable / extendable conveyors rather than taking advantage of the use of forklifts.

❏ Guards or electrical wiring are defective on pinch points and junction boxes.

Ergonomics

❏ The workstation that the worker is using is too low or too high for safe production.

❏ Lighting at the work area is in need of improvement: more lighting is needed, the placement of the lighting has to be modified for the workstation or the lighting repairs are in order.

❏ The tools being used for the job are not ergonomically correct.

❏ Chairs, tables, picking bins, parts baskets and word processing stations are not set to accommodate the size and shape of the worker.

❏ Corners, protrusions, pipes, valves and other obstacles or impediments to production and worker need correcting / movement.

❏ Cartons, boxes, and parts are too heavy to be manually handled. There is a need to reduce the weight of the carton or box to promote safer lifting at the worksite and for the workers at the receiving end of the container.

❏ There is a need to install a hoist or lift at the workstation to prevent lifting and bending.

❏ The worker is using an improper stance and method to lift product or parts.

❏ The design of the workstation requires the worker to use an awkward posture, to lift in front or overhead, to have to bend the wrists, back or neck or to kneel or squat while performing the job.

❑ The grip required to hold and use a tool, or to lift parts and product must be improved.

❑ The job requires numerous repetitions, even with a light force, and must be modified.

❑ The job involves pushing or pulling heavy boxes; a forklift is needed to move the loads.

❑ The tool or process involves excessive vibration and must be cushioned or modified.

❑ The worker is exposed to too much heat, cold or humidity while performing the job.

❑ The job requires modification to protect the worker such as special protective clothing or cooling processes.

❑ The working / walking surface at the workstation need comfort mats.

Dock Safety

❑ Wheel chocks are not being used on trucks and trailers.

❑ A fixed ladder or stairs is not available to go down to and up from the dock well.

❑ Auxiliary lighting, which is used for the trailers, is in need of repair or the lights must be installed.

❑ Storage of product at the dock area is precarious and must be corrected.

❑ A spill clean-up kit is needed at the dock area.

❑ Protective railings are needed for visitors and workers to protect them from lift truck traffic.

❑ Floors of trailers are not being inspected prior to entering a trailer with a forklift.

❑ Wheel chock signs are not available or are in need of repair or replacement.

❑ Automatic trailer restraints are in need of maintenance.

❑ Housekeeping under the dock plates requires attention.

❑ The fire extinguishers at the dock need refilling or inspection.

❑ The workers or visitors at the dock need personal protective equipment.

❑ Dock plates are in need of repair or the capacity is inadequate to support the weight of the forklift and the load.

❑ Electrical wiring or installation is defective and does not meet the National Electric Code.

Ladders

❑ Rolling ladders do not have rubber stoppers on them.

❑ Rolling ladders have visible damage to the side rails, supports, or stairs.

❑ Aluminum ladders are being used near electrical wires.

❑ Portable ladders are damaged and are in need of repair or replacement.

❏ Tools or parts are being left on portable ladders.

❏ Portable ladders are being used at the incorrect angle.

❏ Ladders are being used in close proximity to moving vehicles and could easily be struck by them.

❏ Fixed ladders are damaged and are on need of repair.

❏ The step-off area for a fixed ladder lacks protective railing or a small platform.

Battery Charging / Battery Changing

❏ Acid neutralizer is not available at the battery charging area.

❏ Guidelines on battery maintenance and recharging, supplied by the manufacturer, are not available.

❏ A dry chemical or CO^2 fire extinguisher is not readily available or needs recharging at this area.

❏ Personal protective equipment needed for protection from battery acid is not available. This would include rubber gloves, rubber apron and a face shield.

❏ Workers have not been trained in the safe procedures for battery care.

❏ A mechanical means of pulling the battery from the forklift is not available.

❏ Batteries show signs of corrosion and need cleaning.

❏ No smoking signs are not posted as required.

❏ An eye washing station that provides at least 15 minutes of running water is not present.

❏ Each battery charger has not been labeled for identification in the event of an emergency and the need to disconnect them.

❏ Circuit breaker boxes are not labeled properly to allow for lockout procedures in the event of an emergency or a need for maintenance.

Propane Tanks

❏ Workers are not lifting portable propane tanks properly and risk a back injury.

❏ The "no smoking" sign that is required by the portable propane tank storage area is missing.

❏ The propane storage area is not ventilated properly.

❏ The "O" ring in the propane feed line between the engine and portable tank is missing.

❏ Empty propane tanks are being left throughout the plant rather than being returned to the storage area.

❏ The propane gas line at the portable tank is in need of replacement.

❑ Workers are not wearing heavy duty gloves or eye protection while changing portable tanks.

❑ Workers are not being trained to change portable propane tanks.

Confined Space Entry

❑ The air in the confined space is not being tested prior to entry.

❑ Workers have not been trained to enter confined spaces.

❑ A means of lowering workers into a confined space to rescue workers has not been established.

❑ A rescue harness is not available for conducting worker rescue.

❑ There is no written confined space entry program.

❑ The instruments needed for testing the air in a confined space are not available.

Walking and Working Surfaces

❑ The floor surface has ruts and cracks in it which could cause a fall.

❑ Housekeeping practices are not being observed in the area of the JHA assignment.

❑ Yellow walkway lines are not present at the working area.

❑ Signs are not mounted to warn or alert workers of traffic or other walking hazards.

❑ Lighting at the working area is inadequate.

❑ Hand railing is missing from the working area.

❑ Stairs are at an incorrect pitch and are difficult to use.

❑ Walking surfaces are slippery with oil, grease, sand, pellets, ice, hydraulic fluids, or snow.

❑ Nosings of stairs do not have non-skid surfaces.

❑ Exits and / or non-exits are not identified properly.

❑ Stair treads or hand railings are loose and need anchoring.

Powered Industrial Trucks

❑ Forklifts and other powered industrial trucks have not been inspected before being placed into use.

❑ Seatbelts are not installed on sit-down (counter balance) forklifts.

❑ The equipment lacks the preventive maintenance procedures required by the manufacturers.

❑ Operators are not following the safe operating instructions required by OSHA.

❑ Additional safety equipment such as back-up alarms and rotating warning lights are not working or are not installed on vehicles.

❑ Convex mirrors are not available in areas where visibility is restricted.

❑ Barriers, such as posts, rails, and high-impact cables need to be installed to protect pedestrians from powered industrial trucks.

❑ Lighting is insufficient for safe passage of forklifts and protection of pedestrians.

❑ Operators are not using wheel chocks on trucks or trailers to prevent inadvertent movement.

❑ Operators are not wearing personal protective equipment as required.

❑ Powered industrial trucks that operate out-of-doors and in trailers are not equipped with lights.

Welding

❑ Welders are not wearing leather gloves, helmets, leather jackets, sleeves, or safety glasses while performing the work.

❑ Standing water is present on the floor and working area where welding is taking place.

❑ Maintenance on the welding equipment is not being performed according to the manufacturer's guidelines.

❑ The proper lens shade is not being used in welding helmets.

❑ Open electrical hazards exist on welding machines.

❑ Cylinders on MIG welding machines are not supported to prevent them from falling over.

❑ MIG welding machines are not equipped with wheels that are functional.

❑ Fixed welding stations are not equipped with working ventilation hoses and spouts.

❑ Toxic metals are not being eliminated in the welding production process.

❑ Welding stations are not being adjusted for the height of the worker to prevent back fatigue.

❑ Welding screens are not in place to protect co-workers.

Shelving / Racking

❑ Sections of shelving or racking are bent, twisted, or distorted and must be replaced.

❑ The racking or shelving has not been designed to take the loads that are being placed on them.

❑ Protective posts to protect the ends of racking are missing or are ineffective.

❑ Product is being placed too close to heaters, gas pipes, or sprinkler heads.

❑ Product being stored on upper sections of racking is not shrink-wrapped or wrapped in strapping or banding.

❑ Signs are not available on the sections of rack or shelving to alert everyone of the capacity.

❑ The distance between fixed racking or shelving is inadequate for safety as well as production.

❑ Workers are being endangered from the movement of powered industrial trucks while they are working near fixed racking or shelving.

❑ Racking and shelving is not being supported to prevent toppling in the event of earthquake or being struck by powered industrial trucks.

Working From Heights and Elevations

❑ Fixed railing at least 42 inches high is not in place where production or work is being performed.

❑ Ladders used for gaining access to elevations are not safely anchored in place, are in need of repair or do not extend 3 feet above the step-off landing.

❑ Scaffolding or access ladders are placed in areas where they can easily be struck by passing equipment.

❑ Barricades, warning lights, and signs are not in place on the floor around areas where elevated work is being performed.

❑ Harnesses, lanyards, and other safety devices to prevent falls are not being used as required.

❑ Workers have not been trained to inspect and wear safety harnesses to prevent falls.

❑ Safety belts, not safety harnesses, are being worn which could cause injury to the worker in the event of a fall.

❑ Lanyards being used are too long and could be the cause of worker injury in the event of a fall.

❑ Safety harnesses show signs of wear and have defective parts that need to be replaced.

❑ The working area does not have a means of anchoring the lanyard and snap.

Power tools and Hand Tools

❑ Power tools and hand tools are not being inspected on a regular basis.

❑ Electrical power tools do not have grounding or double insulation to prevent shock.

❑ Power tools or hand tools are not being used correctly per manufacturer's guidelines.

❑ Workers have not been trained in the correct use and safe use of tools.

❑ Personal protective equipment is not being worn as required while using power or hand tools.

❑ Tools are not being stored or maintained properly, i.e. frayed cords.

❑ Incorrect tools are being substituted for the correct tools while performing the job.

❑ Guards are missing from tools that require guarding.

❑ Workers are using awkward positions while using tools and ergonomic improvements are necessary.

❑ Tools such as chisels and drill bits need grinding and dressing.

❑ Ground fault interrupters are not being used in wet or damp areas.

Mezzanines

❑ There are open sections around the perimeter of the mezzanine where hand railing is missing.

❑ Product is stored too close to sprinkler heads at various levels on the mezzanine.

❑ The mechanical lift for product movement requires maintenance or inspection per the manufacturers guidelines.

❑ Signs that identify floor loading per square foot are needed.

❑ Shelves are being overloaded with product.

❑ Exit signs and directional signs are missing from key spots on the mezzanine.

❑ Grating on the mezzanine floors is loose, curled, or missing.

❑ Fire extinguishers are not mounted and identified on the various mezzanine levels.

❑ Stairs being used to access the various mezzanine levels are loose, missing, or defective.

❑ Electrical panels for the mezzanine are not identified properly or are not meeting the electrical code.

❑ Workers are not using correct lifting procedures while manually handling parts and cartons.

❑ Personal protective equipment is not being worn as required.

❑ Emergency procedures are not posted on the various levels of the mezzanine.

❑ Chemicals are not being stored safely or handled safely by the workers.

❑ Low overhead clearance hazards are not identified to prevent a struck-against injury.

Airborne Hazards

❑ Ventilators for capture gases, welding fumes, vapors, dusts, and mists are not working or must be installed.

❑ Personal protective equipment to protect workers from airborne hazards is not being worn or required.

❑ The permissible exposure limits established by OSHA for various chemicals, etc. have not been determined through air sampling in the plant

❑ Management has not replaced highly toxic materials with less hazardous ones in the plant or at the worksite

❑ Professionals, whose job is to detect, measure, and evaluate industrial hygiene hazards, have not completed any plant surveys.

❑ Engineering technology to protect workers from airborne hazards has not been implemented.

Chemical Hazards

❑ Chemicals that can burn, explode, or are combustible are not being handled, stored, or used safely.

❑ Management has not replaced hightly toxic materials with less toxic chemicals.

❑ Workers are not being protected from chemical exposures by the use of personal protective equipment.

❑ Lids are being left off of chemical containers and the chemicals are allowed to evaporate into the work area.

❑ Chemicals lack the appropriate NFPA 704 or HMIS labels.

❑ Material safety data sheets are not available for each chemical.

❑ Flammables such as aerosol cans and small containers are not being stored in fire safety cabinets.

❑ Unstable materials are not stored safely according to manufacturers' guidelines.

❑ There is no spill clean-up kit where chemicals are being handled or stored.

❑ Workers are not trained to safely handle the chemicals they are working with.

❑ There is no written hazard communication program available.

❑ There is no comprehensive list of all of the chemicals in the facility.

❑ Management has not developed a program to request material safety data sheets from vendors during the purchasing process.

❑ There is no emergency eye-wash station or drench-shower near chemical handling or storage processes.

❑ There has been no air monitoring conducted in the workplace.

❑ Workers are not washing their hands before eating or drinking.

❑ Flammables are not being stored in flammable liquid cans that have flame arrestors and self closing lids.

❑ Workers are storing and consuming food in the chemical storage and processing areas.

❑ Chemical waste is not being properly disposed of according to EPA regulations.

❑ Workers are wearing clothing contaminated with workplace chemicals.

❑ Warning signs are not posted to warn employees and visitors of the hazards associated with the chemicals in the work area.

Noise

❑ Noise levels have not been evaluated to determine whether or not hearing protection is needed.

❑ Workers are wearing the incorrect earplugs or earmuffs; they are wearing the equipment incorrectly; they are wearing unsanitary equipment; or they have no equipment whatsoever.

❏ New employees are not receiving baseline audiograms.

❏ Workers in areas with sound levels over 85 decibels are not receiving annual hearing tests and evaluations.

❏ Barriers and structures to absorb noise are not in place or are in need of repair.

❏ Workers have not been trained in the basics of preventing hearing loss.

❏ Management has not developed a written program for the prevention of hearing loss in accordance with OSHA regulations that have to be adhered to.

Summary

Supervisors and employees should have knowledge of how workplace injuries and illnesses occur. The only way to prevent injuries is to eliminate all of the contributing factors. When performing a JHA, the supervisor should have a grasp of what constitutes both unsafe conditions and unsafe behavior. To identify similar situations in the workplace, supervisors should use the descriptions of each provided in this chapter. Causes of injuries include defective or otherwise unsafe conditions of tools, equipment, machines, facilities, atmospheres, actions of management, and workers.

For too long there has been a belief that the worker caused eight out of ten injuries in the workplace. Workers do injure themselves or others by not following prescribed safety rules, but an analysis of typical workplace injuries would indicate that management plays a big role in controlling what goes on in the workplace. Management must properly train workers, provide and enforce PPE, place guards on machines, and otherwise provide a working safety and health program. Once hazards are recognized, management should step in and correct the conditions or behavior before someone is harmed.

Chapter 6

Sources of Injuries

Despite the several millions of injuries and illnesses that occur every year in the workplace, there are only 11 actual ways in which these injuries and illnesses can occur. When performing a JHA, supervisors should evaluate the workplace for these potential sources of injuries and illnesses. In this chapter, each of these 11 sources of injury will be identified along with some typical injuries that are associated with each. Recommendations for preventing injuries are provided for each hazard source. The simulated situations depicted in the photographs illustrate the sources of injury for which supervisors should be looking.

The 11 sources of injuries and codes used to indicate each are:

1. Struck-by (SB)
2. Struck-against (SA)
3. Caught-between (CB)
4. Contact-with (CW)
5. Contacted-by (CBy)
6. Caught-on (CO)
7. Caught-in (CI)
8. Fall same level (FS)
9. Fall to below (FB)
10. Overexertion (O)
11. Exposure (E)

Supervisors should evaluate the work area for each of these 11 sources of injury or illness. When completing a written JHA form, use the shortcut code letters to easily reference the source of potential hazards found on the job. By indicating the type of hazard for each item on the written JHA hazard list, supervisors can better evaluate what type of training the employee needs. Also, the supervisor can use these written notes to check that he has looked for each source of injury and did not miss any.

Classification of Injuries

Struck-by

The "struck-by" injury is one in which a person is struck by an object of some kind. The force can be substantial and abrupt. The person could either be in motion or stationary, but the object that strikes the person must be in motion. The individual preparing the JHA should realize that some objects in the worker's environment are expected to be moving and thus prepare the JHA in a manner that recognizes this. On the other hand, there are some objects that are not supposed to move. If they do, the worker may not be prepared for the sudden movement.

Examples of Struck-By Incidents:

- A worker is walking in a warehouse aisle when a stock chaser makes a sudden sharp turn at a corner and strikes the pedestrian. Figure 6.1 illustrates a simulated situation where a powered industrial machine can strike a pedestrian.

Figure 6.1 Pedestrian in the path of a stock chaser
This photo simulates a situation in which a worker is about to be struck by a stock chaser.

- A worker is using a drill press and has replaced a drill bit in the chuck but has forgotten to remove the chuck key. He turns on the drill press and the chuck key flies out of the machine and strikes him in the head.

- A supervisor is talking to a worker near a punch press coil feed. The coil has just been placed on the spindle and the operator continues to cut the steel strapping holding the coil together. The end of the steel coil whips downward and strikes the supervisor on the calf of his leg.

- A stonemason is using a hammer and chisel to smooth out some used bricks. The chisel is placed on the brick, the hammer strikes the chisel, and a piece of brick strikes the worker just above his unprotected right eye.

Figure 6.2 Box about to fall from a shelf
This heavy carton could easily fall and strike someone.

- A visitor is being escorted through a distribution center and is walking under an overhead conveyor. A few parcels jam on the conveyor and fall to below because the netting is not in place at that spot. The visitor is struck on the shoulder by a heavy box. Product falling from storage shelves is not unusual. Figure 6.2 shows a photo of such a hazard.

- As a final example of a very common struck-by incident, Figure 6.3 demonstrates the potential of being struck on the fingers while hammering a nail.

In each of the above situations, the persons involved did not expect the incident to occur. Here is a short list of things that are not expected to be in motion:

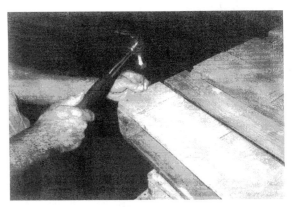

Figure 6.3 Striking a nail with a hammer
When anyone is striking a nail with a hammer, the fingers can easily be struck.

- A pickup truck in a parking lot drifts backward and almost hits a worker who is walking behind it. The truck collides with another vehicle. The driver of the runaway vehicle failed to place the manual transmission in gear and did not set the parking brake.

- Several workers are walking by a scaffold at a construction site when the wind causes several 2x12 planks to fall off of the scaffold. The workers narrowly escape serious injury, but the circumstances could have been much more serious.

- A maintenance worker is repairing an overhead crane runway wheel. A crane in the same bay strikes the crane being repaired, causing the maintenance worker to fall 20 feet. His partner failed to lock and tag the second crane and failed to place warning flags on the bridge and hooks of each crane.

When evaluating a job, always consider safeguards for those things that are not supposed to move but can unexpectedly do so. Supervisors should be alert at all times to recognize and correct unsafe conditions that could cause struck-by injuries. The following potential hazards should be reviewed at each worksite to prevent struck-by incidents and injuries:

1. Where possible, provide fixed walkways with lines and protective barriers as a means of protecting pedestrians from moving vehicles. Yellow lines on the floor and signs at entrance doors and at key locations of the plant can help to warn pedestrians—especially visitors.

2. Improve facility lighting as well as overall visibility. Replace defective overhead lights and replace defective bulbs where necessary. In some cases, lights and reflector shields may need a good cleaning to allow for greater reflection.

   Where blind corners and similar potentially hazardous areas exist, install convex mirrors so that equipment operators and pedestrians can avoid collisions.

3. Install guards on machines where workers could be struck by fixed moving parts or objects being ejected from the machine as they are being produced.

4. Require rigid enforcement of lockout/tag out requirements.

5. Anchor objects or enclose them to prevent movement. As an example, place shrink-wrap on pallet loads of loose product to prevent it from falling from shelving and racking.

6. Require the appropriate alarms and flashing lights on moving machines. As an example, an alarm sounds and a flashing light warns travelers at air terminals that the luggage conveyor is about to move.

7. Use only authorized and qualified operators on powered equipment or machines.

8. Provide more clearance at passageways and corners to prevent struck-by injuries caused by powered industrial trucks or fixed machinery. Use safety 'lookouts' when work is being performed overhead or in tight corners where equipment may be in motion.

9. Additional warnings of hazards could include: flags, back-up alarms, horns, tags, red lights, blinking or flashing lights, and buzzers. Keep alarms functional and keep signs clean and legible.

10. Provide preventive maintenance as required by the manufacturers of the equipment.

11. Train all workers to assume that things can go wrong and move unexpectedly and cause injury. Keep a safe distance from moving objects. Where stationary equipment or product is stored, always be aware of potential movement if the product or equipment is struck by a crane hook or other object, or if someone fails to secure an object.

Figure 6.4 Pipe falling toward worker
This worker is putting up his hand to keep from being struck by a falling pipe.

12. Where possible, re-route traffic: create new traffic patterns to provide a greater margin of safety.

13. Secure objects that are being transported by forklifts, cranes, conveyors, etc. A quick stop could easily dislodge an object off of the forks, crane sling, or conveyor belt. When odd-shaped or unusual-sized objects are being transported, travel slowly and sound an alarm to warn others. Use a spotter if necessary. Never overload a piece of equipment, and place load-limit signs on machines to alert operators.

14. Use the correct equipment or tool for the job at hand. If a load weighs 3500 pounds and a 3000-pound capacity forklift is all that is available, reduce the weight of the load or obtain a larger forklift for safer handling. Never add counterweight.

15. Enforce the use of personal protective equipment (PPE).

16. Use only the correct locations for product storage. If racking is too weak for heavy loads, do not allow operators to place heavy loads on the rack. Many struck-by incidents are caused by falling or shifting materials that are poorly stored, stacked, or positioned. Parts in storage can fall unexpectedly, like the pipe shown in Figure 6.4, so workers have to be prepared for such movement.

17. Use observers or a spotter in situations where visibility is limited or the hazard calls for it. Signs are helpful but will not prevent someone from proceeding into an area where there is a possible hazard.

Struck-Against

A struck-against incident is one in which the worker unexpectedly and forcefully makes contact with something in the worker's environment. Most of the time, the worker is in motion and strikes against a fixed object. As an example, it is not unusual to see someone strike his or her shin on an open desk drawer. While conducting a JHA, look for the potential of a worker to place himself in a strike-against situation. For example, the worker in Figure 6.5 is about to strike his arm against nails that are protruding from broken pallet boards.

Figure 6.5 Protruding nails
This worker could easily strike his arm against the nails in the boards.

Examples of Strike-Against Incidents:

- A construction worker is attempting to loosen a large nut on a steel column. For added leverage, he slides a long section of pipe on the handle of the wrench. He begins pulling on the pipe, the pipe slips off of the wrench, and he tumbles rearward and strikes the back of his head on a handrail.
- A forklift operator parks his truck near a doorway and leaves the forks at knee height instead of lowering them to the floor. A co-worker walks into the elevated forks and seriously injuries his kneecap.
- A warehouseman is loading a trailer while using a portable dock plate to place a load in the trailer. The forks are too low and the tips of the forks make contact with the edge of the dock plate. The operator is propelled forward and strikes his head on the overhead guard of the forklift.
- A mechanic is removing rusted bolts from a muffler on a car. He pulls on the ratchet handle, the nut snaps off of the bolt, and he strikes his hand on a rusted fender.
- A secretary is transferring hand towels to a cabinet in the kitchen of the plant. She places the box of towels on the floor, lifts some towels, and opens the cabinet door to put them away. She bends over to get more towels, straightens up, and strikes her head on the open door.

Struck-against incidents can be reduced or minimized. The solutions can require physical change, a procedural change, or the use of signs or warnings. Struck-against incidents are very common throughout industry. Keep in mind that the worker could be traveling in a vehicle, such as a forklift, or walking, bending, pulling, or pushing on an object. The following is a listing of items that can help in preventing struck-against incidents:

1. If pulling or pushing on a wrench or tool, expect it to slip. The jaws of the wrench should be facing the nut. Take a correct stance to apply pressure to the wrench in such a manner as to not be at risk of losing your balance. Plan ahead regarding the area that the hands or other body parts would strike if the wrench or tool slips. Use the other hand to stabilize your body and limit rebound. Never use "cheater" bars for leverage; use heat or rust removers that are intended to loosen the nuts.

2. Inspect the facility for the purpose of seating, pushing back, or adding guards to any objects protruding into aisles, work areas, or walking / working zones. Where doorways or doors open into rack storage holders for pipe, bars, or rods, place a fixed barrier at the end of the rack so none of the parts would protrude into the walking area.

3. Keep work areas well-lighted so that hazards can be detected more readily.

4. Enforce the use of personal protective equipment. As an example, hard hats can protect a head when a worker stands up while working under racking. It is not unusual for a worker to bend over into racking to retrieve parts and product. In some cases the workplace may not be changeable; therefore, PPE would be necessary. In Figure 6.6, the worker is about to strike his hard hat against the underside of racking.

Figure 6.6 Striking head against bottom of rack
Hardhats will protect the head from striking against the racking as shown in the photo.

5. Require workers to follow safe procedures when having to exert force or pressure on an object. Brute force and strength should only be used as a last resort. Heat, rust removers, and the use of a hammer to tap a nut or bolt loose are recommended.

6. Overall housekeeping should be at the high quality level at all times. This care and attention to the workplace will help reduce struck-against incidents.

7. Use warning signs, lights, or rope off areas that could contribute to struck-against incidents. Identify low-access areas that may need painting to highlight a hazard. Add guarding or padding where workers must duck under beams, conveyors, etc.

8. Train powered industrial truck operators to be alert for objects in their path of travel. Because of the speed in which they are traveling, they might not see an obstacle in time to avoid running into it.

9. Always remind workers to look before they walk. Pedestrians are to be observant of areas in front of them from the floor to above the head. A protruding object could easily be lying in wait for the unsuspecting pedestrian.

10. Relocate or remove valves or other fixtures that may be located in walking zones. Allow for the inspection of new machines prior to purchase. Where possible, apply the "safety through design" principle and design the hazards out of the process in advance.

11. Yellow paint is very effective at highlighting an obstacle. Yellow and black stripes painted on columns, crane hook blocks, protective posts, overhead doorframes, and machine guards will provide greater visibility.

12. If possible, re-position machines or operator workstations to alleviate tight quarters, which could easily be the cause of struck-against incidents. Where employees are working in tight quarters, a barrier of some kind may be needed to prevent injury.

13. Always use the correct tool for the job to prevent slipping or movement of any kind.

14. Train all workers and supervisors to be aware of any items or situations that could produce a struck-against injury. Figure 6.7 provides a graphic example of a

Figure 6.7 Splintered board on a pallet
This sharp sliver of wood is a serious hazard if walked into.

struck-against injury that is about to happen. The worker is about to walk into a very sharp piece of wood protruding from a broken pallet board located at chest height.

Caught-Between

When considering what a caught-between incident is, always think of pinch points. Most workers are familiar with this term because injuries that produce caught-between injuries are very common throughout industry and can involve a finger, hand, arm, foot, leg, torso, or the entire body. Pinch points are usually easy to identify. When performing a JHA, look for pinch point hazards and make the employee aware of them at that time.

Examples of Caught-Between Incidents:

- An employee in a warehouse is standing at a dock door directing a trailer that is backing. His foot is on the edge of the dock while he leans out to give hand signals to the driver. The trailer makes contact with the dock bumpers and, in the process, the employee's foot is caught between the dock and the trailer.

- A laborer in a brickyard asks a forklift operator to lift him 10 feet in the air on the forks because he needs to change a light bulb in the work area. While standing on the forks and facing the operator, he places his hands on the mast for balance. As he is being raised, the small finger of his right hand is severed between a roller in the mast and the forklift mast frame.

- A forklift operator on a stand-up lift truck has just set a pallet on a section of racking. While backing, he fails to keep both feet in the operator's compartment. His right foot is outside of the framework of the truck. He backs into a large concrete-filled pipe that is used as a racking protector. His right heel is crushed between the lift truck and pipe.

- A construction worker is stacking large concrete blocks while standing on a scaffold. As he is placing the blocks on the scaffold planks, he lowers a heavy block on his right hand, trapping it between the lower block and the one in his hand.

• A maintenance worker is trouble-shooting a machine. To maintain his balance he places his left hand on a metal rod used as a slide on a machine. To observe the problem with the machine he pushes the start button with his right hand. The machine cycles, trapping his left hand between a part of the machine and the moving machine part.

Figure 6.8 depicts a very typical caught–between incident, fingers being caught in a door jamb.

• The simulated situation in Figure 6.9 is an example of a backing forklift pinning a worker to a fixed object. In this case, the worker is about to be pinned against a pallet load of product.

Pinch points can be found where there are normally moving objects, such as moving parts of a machine, mobile equipment, or manual handling of parts and product. A good example of such an exposure can be seen in Figure 6.10.

The worker has placed his hand on the reach mechanism of a narrow-aisle forklift. If the forks are retracted, the worker could lose his thumb. A caught-between incident can also occur if the worker is pinched or crushed between a fixed and moving object, or two moving objects. As with injuries that are caused by struck-against exposures, caught-between injuries can also occur in the following situations:

• If a worker is in the path of a normally moving part and a stationary part;

• If a worker is in the path of a normally moving part that deviates from its path of travel; and

• If the worker is in the way of a normally moving object that opens or closes unexpectedly.

As mentioned earlier, many workers frequently use the term "pinch point" to identify a specific hazard. This awareness is a giant step in reduc-

Figure 6.8 Hand caught in a door
This photo is demonstrating a very common caught-between injury – the hand trapped in between the door and frame.

Figure 6.9 Backing forklift
This backing narrow–aisle lift truck can easily trap the worker between the pallet and the truck.

Figure 6.10 Hand placed on reach mechanism on forklift
If someone was to place his or her hand on the scissors portion of a reach type of forklift, the injury would be severe.

ing caught-between injuries. Of course, just being aware of a hazard does not always provide protection for the worker; protection must go beyond the awareness level. The following items can help to prevent "pinch point" hazards:

1. Place guards on and around moving parts of machines and processes. In many cases, guards are placed on machines by the manufacturers, but are left off during maintenance or repair in the plant. If the plant has a low level of safety awareness, guards can easily be left off or discarded. This action can easily pose a hazard to workers. A typical example of a machine that is constantly in need of guarding (as well as attention) is the common bench grinder. In order to change the worn stone, the guards have to be removed. Data from OSHA inspections indicate that citations for unguarded bench grinders lead the list for violations of the machine-guarding standard. It is not unusual for the side guards and tool rest to be left off after new stones have been replaced.

    The best way to identify pinch points on machines and other operations is to search for them. Management may have to be creative to develop guarding on old machines. Seek the advice and assistance of workers on those machines and processes. In addition, look at past records of injuries and incidents to identify sources of caught-between claims from workers' compensation loss runs.

2. Highlight hazardous areas. Yellow paint is usually painted on the outside of machine guards. The inside of guards should be painted orange to provide highlighting when the guard is open. Yellow and black striped tape can be placed on moving or stationary parts to remind everyone that a pinch hazard exists.

Figure 6.11 Pulley belt without a guard
The guard for this pulley belt drive is missing and a replacement guard is needed.

3. Improve lighting where necessary to help workers identify pinch hazards on their machines or at their jobs. Many times machine parts are all black in color and hazards are not easy to identify.

4. Instruct maintenance personnel to always replace machine guards. It is not uncommon for guards to be left off of a machine for one reason or another. Also, ensure that all screws bolts, nuts, and fasteners are used when replacing the guard. Air compressors are machines that usually contain a partial or missing guard because someone has performed maintenance on the machine and failed to replace the guards. In Figure 6.11, a portion of a guard has been left off of the pulley belt and poses a hazard to anyone near the pinch point.

5. Everyone must use the lockout / tagout rule when performing maintenance or repair on any machine. Many times a machine will cycle or activate while someone is working on it. If the machine's movement is unexpected, any part of a worker's body could be caught between two objects. To prevent any unexpected or unwanted movement, the machine must be brought to a zero-energy state. Either bleed off stored energy or fluids, block out a part of the machine that could inadvertently

move, or remove the electrical power feeding the machine. Use only qualified electricians to perform this type of work.

6. Design facilities, departments, or machines in such a way that the risk of injury from caught-between incidents is zero. Study plant layouts in advance and design safety into the process. It makes good business sense to eliminate a problem during the design phase of the project rather than after a worker has already been injured.

7. Encourage workers to search for and have pinch points and other hazards corrected. Safety teams or safety committees can offer a lot of assistance on this subject.

8. Require specific workplace attire. Occasionally, the newspapers carry a story about a worker that pulled into a machine as a result of unsafe work clothing. Other causes of these types of incidents are long hair, jewelry, or loose clothing.

9. Where machines, such as automotive hoists, move in a stationary up and down motion, paint a boundary line on the floor to keep workers from the motions of the machine. It is not uncommon for someone to have their feet trapped between a hoist that is being lowered and the floor of the garage.

10. Use flashing lights or alarms to warn everyone that something is in motion. A back-up alarm on a forklift can help in the prevention of someone being caught between a solid object and the backing forklift.

11. Secure loads with chains, strapping, webbing, or banding to prevent movement. Expect product to move or shift if it is not anchored in place.

12. Place signs reading "pinch point" where these hazards are present. Ensure that signage meets American National Standards Institute (ANSI) guidelines.

Contact With

Injuries associated with "contact-with" incidents mostly involve someone making contact with something hot, a chemical, or electricity. The contact is usually non-forceful. Contact with chemicals is one of the major causes of dermatitis in the workplace. OSHA`s Hazard Communication Standard recognizes the potential for contact with chemicals and provides a guide for ensuring that workers are aware of the hazards in their area.

Figure 6.12 Sharp edge of manufactured pipe
The end of this pipe is razor sharp and a hand or finger could easily come in contact with the sharp edge.

Additional contact-with exposures exist when the worker makes contact with something sharp or something sharp that is in motion. In Figures 6.12 and 6.13, workers are potentially exposed to such hazards. The worker in Figure 6.12 has discovered a razor-sharp edge on this piece of manufactured pipe. Anyone moving their hand toward the pipe could make contact with the sharp edge. This type of exposure can also fit into the struck-against category of hazard. In Figure 6.13, the board being pushed toward the rotating band saw blade can cause an injury. As the board is being pushed the hand can make contact with the blade. Again, some may also, (correctly), classify this as a struck-against injury. In both of these examples, little or no force is exerted on the part of the person when making contact.

Figure 6.13 Cutting a board on a band saw
The workers hand could easily make contact with the moving band saw blade. Also, the height adjustment needs to be lowered on the blade.

Typical Contact-With Incidents:

- An auto mechanic has just used a cutting torch to remove a muffler clamp, and it falls to the floor. Several minutes later, he forgets that the part is hot, picks it up with his bare hand, and suffers a burn to his hand.

- A maintenance worker is changing a battery in a truck and the terminals and cables are heavily corroded. He works without gloves and removes the battery. While installing the new battery, he forgets that his bare hands have made contact with the corrosive. With the corrosion on his hands he rubs his eye and causes an eye injury.

- A warehouse worker drops a carton of chemicals that are contained in plastic bottles. The box hits the floor and some of the bottles break open. He reaches for some rags to wipe up the spilled liquid and his hands make contact with the chemical. The simulated incident in Figure 6.14 identifies a worker that is about to touch a spilled chemical.

- A worker is using a defective extension cord and trouble light while working on a machine. His ungloved hand makes contact with a bare wire and he receives a severe shock.

Figure 6.14 Touching a spilled chemical
In this photo simulation, the worker is about to touch a spilled chemical without the correct gloves.

- A construction worker stumbles on some pallets near a large drum that is being used to provide heat at the jobsite. The drum is very hot. He stumbles forward and he puts his hands out to break his fall. His hands make contact with the hot drum and he burns his hands.

- A worker touches a live electrical wire. In Figure 6.15, a supervisor has discovered a damaged electrical box that could cause a contact-with incident of this type.

• The worker in Figure 6-16 has placed his hand in the direction of the box-cutting knife being used. He can easily suffer a serious injury after being cut by the knife blade.

Figure 6.15 Damaged Electrical Box
This damaged electrical box could cause a shock or other serious contact-with injury

Many contact-with incidents startle or surprise the victim. One is not prepared to be shocked by a bare wire or suffer a burn to the hand when touching a hot object. The surprise of the incident may cause the worker to react spontaneously in an unsafe manner. Proper planning and education can help to prevent further injuries and incidents by training the worker in proper response procedures.

Contact-with injuries or incidents can be minimized or reduced by placing a focus on:

1. Requiring workers to wear the correct PPE. In some cases, the use of the correct PPE will prevent secondary injuries. If a worker is burned while handling a chemical or hot part, they could easily drop the hot part into a chemical container or into molten metal. In addition to injuring the hands from touching the hot parts, the injured worker could easily splash their face, neck, or arms after dropping the part. Proper battery maintenance can expose workers to the harmful effects of acid and corrosion.

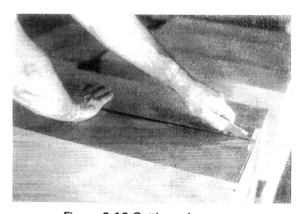

Figure 6.16 Cutting a box open
Never place the hand in the path of a cutting knife.

2. Providing rubber mats or ground-insulating mats to protect workers from potential electrical shock. Even low voltage can cause shocks.

3. Requiring maintenance workers that have the potential of touching live electricity to wear special insulated shoes.

4. Requiring inspections of electrical wires and cords such as extension lights, cords that power electrical tools, and extension cords. Inspect for missing ground prongs, bare wires, damaged cords, and grounding. Install ground fault circuit interrupters where necessary. Ensure that all electrical work complies with the National Electric Code.

5. Providing guardrails and toeboards for anyone walking or working near chemical vats or molten metal.

6. Developing an emergency plan for chemical spills. The plan would include clean up kits, proper disposal methods, PPE, neutralizers, training and follow-up drills.

7. Training workers to search out potential electrical hazards or hot surfaces that could cause injury. Once discovered, correct the hazards.

8. Permitting only trained and authorized workers to make repairs.

9. Where possible, using the lockout / tagout program to keep live parts de-energized. Consider all electrical parts as energized until proven otherwise.

10. Providing guarding or barriers to prevent contact with live current or hot parts.

11. Providing the appropriate signs and warnings to alert others of hot surfaces or hazards associated with hot materials or electricity.

Contacted By

The contacted-by injury is somewhat similar to the contact-with incident. Instead of the worker initiating the action that causes contact with the object, the worker is somehow forced by other means into contact with the object. A worker can be sprayed by an acid, contacted by hot steam or gases, or splashed by a chemical or substance. Most of these incidents involve non-forceful contact. To determine if it is a true "contacted by" injury, the injury must be caused by the injurious characteristics of the contacting agent and not the force of the contact. The injurious agent can be toxic, extremely cold, extremely hot, corrosive, radioactive, electrified or otherwise injurious.

More and more chemicals are being used in retail establishments, warehouses, construction sites and manufacturing plants. As a result of the large quantities of chemicals in the workplace, many workers are injured when they are splashed by chemicals being mixed, applied, being carried, sampled or poured. Injurious liquids can be poured too rapidly, causing a splash to the person doing the pouring or the person working next to him. Again, use OSHA's Hazard Communication Standard as a guide for communicating chemical hazards to workers.

Contacted-By Incidents Include:

• A maintenance worker is repairing a steam line and fails to blank the line or lock out the steam pressure. He applies a wrench to the line and hot steam makes contact with his face and arms.

• An employee in a battery repair shop is heating and melting lead to pour into the battery post mold. He fails to apply enough heat to the mold in advance and is splashed by molten lead while it is being poured because moisture was present.

• A mechanic is carrying a used battery to an area where old batteries are kept. The battery slips out of his hands and falls to the floor. The edge of the battery strikes the floor and splits open. His pants and shoes are covered with battery acid.

• A forklift operator is raising a pallet load of drain cleaning chemicals in boxes onto warehouse racking. He strikes a corner of the high rack and some of the containers fall off of the pallet and strike the overhead guard. The containers break open and he is covered with the liquid chemicals.

• A maintenance worker is cleaning parts in a container. A part slips out of his hands and falls into the cleaning liquid. He is splashed in the face from the liquid.

- The forklift operator in Figure 6.17 is reconnecting a fuel hose on a propane tank and is not wearing eye and face protection. He could easily be contacted by propane gas that has been sprayed in his face. The cigarette is meant to identify another hazard around propane tanks: no smoking!

There are steps that can be taken to prevent contacted-by incidents and injuries from the manual handling of materials, while repairing injurious material containing equipment and from pressure equipment failures.

1. Require the use of specific PPE depending on the job at hand. Consider face shields, rubber aprons, rubber gloves, goggles, safety glasses, and other safety apparel.

2. Require special procedures and policies for those carrying and handling injurious materials.

Figure 6.17 Changing a propane tank
In this photo, the worker is in the process of changing a propane tank. He must not be smoking and should also be wearing a face shield.

3. Require the lockout and tagout system for all maintenance work being performed.

4. Design equipment so that maintenance or repair does not jeopardize workers. Mark all lines and valves so that there is no error while they are being repaired. Replace worn parts and equipment as required by the manufacturer, follow manufacturers recommendations for maintenance and the use of the equipment.

5. Ensure that the correct tools and containers are used for transporting, handling, sampling, brushing, pouring, mixing, and dipping processes.

6. Where possible, avoid the direct handling of dangerous materials. Utilize special equipment or tools for this purpose. Design and develop the equipment to fit the situation.

7. Where possible, use a safer chemical or material in the facility. Work with purchasing to search for chemicals that have higher flashpoints, are less corrosive, and have high numbers for OSHA permissible exposure limits (PEL's) or threshold limit values (TLV's). The properties of chemicals can be found on the material safety data sheets from their manufacturer.

8. Ensure that workers use approved containers and tools as provided.

9. Consider ergonomic issues such as the weight of containers, design of tools, and means of transporting various chemicals.

10. Institute an inspection program for containers and tools. Workers should be required to check the condition of all containers before use and to turn in ones that are defective. Repair minor leaks promptly so that they do not become major leaks.

11. Identify all containers for their content. Label and color-code all pipes and valves so error is minimized or reduced in the event of an emergency or when maintenance is being performed.

12. Develop safe shut off and opening procedures. Workers are to assume safe positions when opening and closing valves.

13. Use special pre-job procedures and instructions to workers assigned to the maintenance of machines.

14. Use only experienced and authorized employees to perform maintenance.

Caught On

The caught-on incident involves a worker having a part of his clothing, working attire, or body caught on a moving or stationary object. It is not uncommon for a worker to snag his pants pockets or cuffs on a nail or other protrusion at home. While at work, this same worker could encounter items such as: ends of steel banding, nails, crane hooks, broken strands on a wire rope, valves, ends of pipe, tails of coils, and boards. Any item that is sharp is most likely to cause an injury.

Something that has been placed in a temporary position can snag a passer-by, especially if the person is unfamiliar with the location. A machine could be temporarily installed near a commonly used walkway. A worker could easily have a loose shirttail caught on a revolving part of the machine. The work area could be dark and a hazardous projection could be hidden. Caught-on hazards are common to all workplaces as well as in the home.

Examples of Injuries From Caught-On Exposures:

- An experienced machinist was operating a large floor drill and while drilling through a thick steel plate, he had to remove the shavings. Instead of using a long-handled brush to remove the shavings, he reached toward the rotating drill bit with his gloved hand. The glove got caught on the shavings and his hand and arm were drawn into the rotating drill. Fortunately he had the presence of mind to turn the drill off with his free hand. However, in the blink of an eye his two middle fingers of his right hand were amputated, he fractured his wrist and dislocated his shoulder.

- A crane hook-up man in a steel service warehouse unhooked a load of angle iron. He signaled the overhead crane operator to lift the hook and chains. He failed to pay attention to his gloved hand and the crane hook. The crane hook caught under his glove at the wrist and lifted him about five feet into the air before the crane operator realized what had happened. He was lowered and only suffered a sore wrist and shoulder.

- A construction worker was dragging a long section of wire rope to a remote part of the jobsite. The loop on the end of the wire rope got caught on a stake in the ground. The sudden stop jerked the worker off of his feet and he injured his back.

- An office salesman was trying to fill a rush order by going out into the warehouse to retrieve a part. This was not a part of his job and he had no business in the warehouse. Instead of using a rolling ladder to gain access to an elevated storage, he climbed up on the racking. He took hold of the part and jumped down off of the rack. His wedding ring caught on a protrusion on the racking and his ring finger of his left hand was torn from his hand.

- A plant manager was called to the shop floor to look at a machine that was operating improperly. The plant manager asked the worker to start the machine so it could be evaluated. The plant manager leaned over to look but failed to realize that his necktie was dangerously close to a rotating gear. The gear caught his tie but the worker was able to shut off the machine. The necktie was destroyed and the plant manager was badly shaken but unhurt.

- In Figure 6.18, the worker has caught his pant cuff on a pipe hook while walking in an aisle.

- In Figure 6.19 a workers foot is caught under the edge of a carpet and could cause a fall.

Much can be done to protect workers from being injured by caught-on hazards:

1. Provide adequate lighting so dangerous parts of machines and protruding objects are visible. Many injuries that involve trips and falls and snagged clothing take place in areas that are poorly lighted.

2. On a regular basis, supervisors and workers should be on the alert for caught-on hazards. There must be a means of training workers to be aware of the exposures that create caught-on dangers and risks. As with any workplace hazard, caught-on hazards should be corrected as soon as they are detected.

3. Add guards to rotating shafts and couplings on all machines. Be sure to check behind and under each machine to ensure that all of the exposures are guarded. An open shaft or coupling could easily grab a worker's loose clothing and pull him into the rotating parts.

4. Paint immoveable parts, valves, protrusions, and guard covers with yellow paint so that the hazard is more easily identified. Paint the inside of guards with an orange color so the open guard could easily be detected and replaced or closed.

5. Where possible, eliminate permanent projections and other items that create caught-on hazards. A valve protruding into a walkway at ankle level could easily cause an injury. Relocate the valve or rotate the machine to reduce the risk. Remind the maintenance department to look for risks such as these and correct them. In Figure 6.20, a worker has caught his shoe on the edge of a rolling ladder that was protruding into an aisle.

Figure 6.18 Pant cuff snagged on pipe bracket
The workers pant cuff has snagged onto a metal bracket.

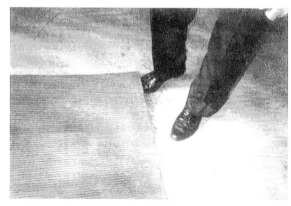

Figure 6.19 Shoe under corner of floor mat
The foot being under a floor mat will usually cause a trip and possibly a fall.

6. Enforce a safe clothing rule. Rings and other jewelry should not be worn while in the plant or at the jobsite. Neckties constitute a serious hazard around moving machinery. Loose soles on shoes, rags hanging from a belt, and loose shirt tails all pose a potential for injury.

7. Train workers to stay clear of moving machinery. A moving conveyor, forklift, or crane all provide the potential for caught-on hazards.

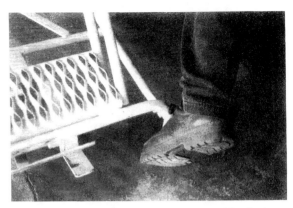

Figure 6.20 Shoe caught under ladder leg
While walking in an aisle, this worker caught the end of his shoe under a rolling ladder leg.

8. Use the lockout / tagout rule when working on machinery. In most cases, it is not necessary to allow a machine to be operational while it is being serviced. Many workers assume that the job will only take a few minutes and risk injury by not locking out.

9. Hand railing and guardrails can provide safety through distance. Where fixed guards cannot be used, workers can be kept away from the hazards by barriers. Yellow walkway lines also help to keep people away from a hazard.

10. Install signs and warnings that indicate where a caught-on hazard may exist.

11. Instruct workers to carry ropes, wires, cables, and wire rope so that they are not injured if a part of the material becomes caught on an immoveable object.

Caught-In

A caught-in incident is one that involves the person or a part of the person's body being caught in an enclosure or hole of some kind. Caught-in incidents do not occur very often. The most common of these limited situations is when a person may place a foot into an open hole or grating and falls. Despite the fact that these types of injuries are uncommon, they can still cause injury or death.

The film industry has created an awareness of caught–in situations that many can relate to or visualize. Films have shown that it is not difficult to lock someone in a safe or vault. The safe may not be opened because of a timer on the lock mechanism. Years ago it was not unusual for children to lock themselves in a refrigerator that had been discarded. It was impossible to open the door from the inside. To prevent these needless tragedies, local governments passed laws that required that refrigerator doors be removed before being discarded.

Examples of Caught-In Incidents:

- A customer is walking through an auto repair shop to look at his car, which is on a hoist. He fails to notice that a section of grating by the garage door is missing. He then steps into the hole, falls forward, and breaks an ankle.

- A worker is welding in a boiler that is unventilated. At the other end of the boiler someone closes the door and the worker becomes trapped and suffocates.

- While attempting to enter a confined space, a maintenance worker becomes stuck in the narrow opening of the pipe he is using for access. Despite the harness and safety line he is attached to, the rescuers have a difficult time getting him out. Fortunately, he is rescued and suffers only minor bruises.

- In an old warehouse, a large vault is used to store items for employee incentive programs such as shirts, caps, jackets and radios. A new supervisor walks into the vault and realizes that there is no means of opening the door if someone is locked inside. Maintenance modifies the safe so there can be no caught-in incidents.

- A warehouseman fails to check the condition of a trailer floor at a dock. He places several pallet loads of product in the trailer. During entry with another load, the floor gives way and the forklift drops to the left side. The hole in the floor has trapped the wheel of the lift truck. As a result of the trapped tire, the operator is propelled forward and strikes his head on the overhead guard.

- Catching a foot in a hole while walking is one of the more common incidents in this category. As an illustration, Figure 6.21 depicts a small hole in an established walkway that is not covered. A worker could easily step into such a hole and his foot becomes trapped. This is a typical caught-in type of incident.

Being that caught-in hazards have the potential to cause serious injury or a fatality, management must do all it can to correct the situations that can cause injury. Caught-in hazards are rather obvious if they are included in a systematic inspection process in the facility. Consider the following measures to prevent caught-in hazards:

Figure 6.21 Hole in floor
A hole is shown in a well.used walkway. A foot would be easily caught.

1. On any door or hatch, place a handle on the inside for emergency exiting. Post appropriate signs on the doors warning of the hazards as well as the potential need for confined space procedures.

2. Check walkways, stairs, and walking surfaces for any gaps in the grating. Replace curled, bent, or deformed grating. Ensure that walkway lighting is adequate to protect the pedestrian.

3. Look for holes in trailers, roadways, and on other surfaces where powered equipment or workers are engaged in work. Patch, fill, and cover all open holes as they are discovered.

4. Place barriers, guardrails, and warning lights and signs by hazards that could trap a foot or cause a part of the body to become caught in an opening.

5. Ensure that lighting is adequate within the facility to alert employees and visitors to caught-in hazards.

6. Train workers for safe entry into tight spaces or confined spaces. Consider the size of the worker in relation to the opening that has to be accessed.

7. Use lockout / tagout procedures to prevent unauthorized access to hazardous areas. Workers have been killed because someone failed to lock or tag a machine or dangerous operation. In one case, two employees were in a debarking machine performing routine maintenance. They failed to lock and tag the disconnect to the machine. The fellow worker that started the machine failed to look inside before starting it. Both workers were killed by the tumbling action of the machine.

8. Follow the OSHA confined space guidelines to prevent caught-in situations.

Fall Same Level

Falls kill and injure many workers each year. In the home, injuries caused by falls are listed as the number one killer. Many falls occur at the walking level or on stairs and steps. All falls have to be regarded as potentially serious. Fortunately, much can be done to prevent falls from occurring.

To fall, the person can be walking, standing, or running. Workers should not be allowed to run in a facility except in an emergency situation. Management must be aware that floor conditions can vary significantly during the course of a day and change with weather conditions. It should be pointed out that many falls occurring at the same level involve the element of chance as to whether or not the fall will result in an injury. A spot of grease or oil on a walkway can pose a hazard to many people, but others will step over and around the slick spot. Those that do fall may or may not incur an injury. Factors such as how the person falls, how naturally coordinated they are, and others could be the difference between a minor and a serious injury.

Examples of Falls at the Same Level:

- A worker is walking to the break area while reading a newspaper. He fails to notice a section of pipe sticking out of a rack at ankle level. He trips on the pipe, falls forward, and breaks his glasses.

- A senior citizen is walking into a retail store and fails to see the curled edge of a carpet. Her foot catches on the carpet edge and she falls on her knees. She is badly shaken but not injured. Edges of carpeting and rugs are capable of producing a fall when the carpet edge is not flat on the floor. The entryway carpet in Figure 6.22 is curled at the edge and could easily cause a trip or fall.

- A visitor to a steel fabricating plant steps onto a four by four foot platform assuming that it is stationary. The platform is equipped with wheels for ergonomic reasons and the visitor falls backwards. He tears his suit but is not injured.

- A forklift has been leaking hydraulic fluid from a hose for several weeks. The operator has not been conducting daily inspections. Management has not been insisting

Figure 6.22 Curled corner of carpet at door
This curled carpet edge could easily cause a trip incident.

on pre-shift inspections so the leak goes uncorrected. A supervisor slips and falls after stepping on the fluid and fractures his elbow.

- An office worker fails to close a file drawer while accessing documents. A worker from the mailroom rounds a corner and fails to see the open drawer. He strikes the drawer, falls forward, and injuries his shin, hands, and arm.

The following steps can be taken to help in the prevention of falls on the same level:

1. Establish a strong preventive maintenance program so that oil, grease, and other fluids are kept off of the floor. As an example, where vehicles are being repaired, antifreeze spilled on the floor can be as slippery as oil.

2. Enforce good housekeeping practices, and clean up spills as they occur. Materials that can contribute to a fall could be granular materials, sugar, leaves, or plastic pellets. Good housekeeping includes keeping the walking and working surfaces as dry as possible.

3. Maintain a policy of cleaning up debris, etc, after completing projects. Put tools and materials back in their original places. Use shelving, racking, tubs, containers, brackets, and pegboards to store work items.

4. Place wires, cables, and cords overhead rather than on walkway levels. Plan the maintenance and engineering plans in advance. Where it is not possible to reroute fixed trip and fall hazards, identify them by painting them yellow and using signs or flashing lights.

5. Institute an inspection process that allows for the identification and correction of trip and fall hazards.

6. Patch holes in pavements, floors, parking lots, and roadways. Seek out holes and irregularities in pavements and floors. Patch floor cracks where possible.

7. Instruct workers regarding the safe way to walk in the facility. For example, hold on to hand railing when using stairs, or step over raised railroad tracks. Steeping over an obstacle includes elevated forks from a fork lift. In Figure 6.23, the worker's foot is under the fork and will surely trip the worker. Figure 6-24 shows that the worker has fallen as a result of the elevated forks.

8. Maintain proper lighting in the facility. Replace defective bulbs and ballasts. Clean bulbs and reflector shades to maximize the lighting.

9. Require that all workers maintain their shoes in a safe condition. Loose soles can easily cause a trip and fall incident.

10. Install non-skid flooring where past incidents or injuries have taken place. Place non-skid strips on stairs and landings. Anchor loose carpeting in place to prevent slips and falls.

Figure 6.23 Elevated forks of forklift
Forklift operators are supposed to lower the forks flat on the floor. This elevated fork could cause a fall.

11. Maintain walking areas when there is snow and ice on the pavement. Use environmentally safe chemicals as melting agents.

Fall To Below

A fall to below incident is one in which at least one person falls from one level where they have been working, standing, or walking to a lower level. Falls to below can occur while walking up stairs, going up a ladder, coming down a ladder, working on a roof, or climbing up to a scaffold, runway, or platform. Falls can also take place to a lower level from the working or walking level. A worker can fall into a hole, into a shaft, off of scaffold planking into a hole, or go off of an edge with a piece of powered equipment such as a forklift.

Figure 6.24 Fallen worker from trip on forklift forks
The worker has tripped and fallen as a result of the elevated forks.

Workers will sometimes take shortcuts that they think are safe because they are in a hurry or because getting a ladder or other device to make the climb safer is not available. For example, the worker in Figure 6.25 had to change a fluorescent light bulb so he elected to climb racking rather than use a ladder. In Figure 6.26, the worker is using a wire cage as a means of getting to the product being counted during inventory rather than using a rolling ladder.

Examples of injuries that can occur from a fall to below:

- A maintenance worker was assigned to paint a wall inside the plant. While painting in a corner, he was several feet short of completing a section to his right side. Rather come down from the ladder and move it so he could easily reach the bare spot, he overreached and fell from the ladder.

- A forklift operator was unloading a trailer at a dock and failed to ensure that the trailer's wheels were properly chocked. While his machine was crossing the dock plate, a landing wheel on the trailer buckled and the trailer lurched forward and to the side. This gap at the dock plate caused

Figure 6.25 Worker climbing racking
Rather than using an extension ladder to reach a light fixture, the worker is climbing the racking.

Figure 3.26 Worker standing on wire basket
This worker is using the edge of a wire cage rather than using a rolling ladder to count inventory.

the forklift to go into the dock well where he was thrown onto the pavement and injured his head and shoulder.

- A bricklayer was working off of a scaffold that was only protected with a handrailing at its front. There was no railing at the ends of the scaffold. These sections were left open for access to the scaffold as well as placement of the bricks needed for the project. As the worker was backing up, his heel caught on a stack of bricks, causing him to fall backwards off of the scaffold. He suffered serious injuries to his back and head.

- At a fast-lube site, customer left his wallet in his jacket inside the car. To get to his car, he had to step over the open pit that workers use to work on cars from beneath. For whatever reason, he fell into the pit and struck his head.

- A maintenance worker placed a long ladder against a column so he could gain access to an overhead light. The ladder was not close enough to the light fixture so he stepped on a large ventilation duct. The duct was not capable of supporting his body weight and he and the metal duct crashed to the floor. He suffered a broken leg.

Because falls to below are life threatening, consider taking the following steps for safeguarding:

1. Cover all open holes with metal plates, plywood, or planking. Anchor floor covers in place. A graphic illustration of the removal of floor hole covers has happened at several construction sites. A large sheet of plywood was used to cover a hole in a walkway. Two workers came along and needed a section of plywood just like the one covering the hole. Both workers picked up the plywood, one at each end. As they stepped forward, the worker at the rear of the plywood fell into the hole they just uncovered.

2. Inspect all portable and fixed ladders for defects. Look at rungs, side rails, and safety devices. Design and install fixed ladders to meet OSHA requirements.

3. Never place portable ladders at doorways or places in the plant where they could be struck by a door or piece of powered equipment.

4. Where workers have to work off of a platform or basket, ensure that each worker is wearing a safety harness. Inspect the workstation and harness before putting them to use. Where it is not practical to use fixed platforms and handrails, a safety net can be used for worker protection.

5. Train workers to identify fall to below hazards. In addition, include training for the wearing of safety harnesses and the erection and care of safety nets. In Figure 6.27, the forklift operator is exposing himself to a fall by not wearing his safety harness and has not lowered the side rails of the working platform.

6. Place barricades, flashing lights, warning signs, and spotters around fall to below hazards to prevent incidents of any kind. Where a hazard is a permanent one, use fixed barricades that cannot be moved.

7. Erect scaffolding per manufacturer's and OSHA guidelines. Cleat or wire all planking so there is no movement of the planking. Provide safe access to the scaffolding so workers are not at risk getting to their job. Anchor scaffolding to fixed structures to prevent collapse and toppling.

8. Improve overall lighting to warn workers of the fall hazards in and out of the facility.

9. Include the search for fall-type hazards on the plant or site's inspection form.

10. Position ladders at correct angles. Use the 4 to 1 rule: for every four feet of a ladder, place the ladder one foot from a wall. Inspect fixed ladders for the quality of the rungs, side rails, and landings.

12. For portable ladders, place non-skid feet on the bottom of the ladder. Where non-skid feet are not available, require that workers place their shoes against the ladder to prevent it from slipping. Wear a harness and tie-off to prevent a fall.

Overexertion

An overexertion incident or injury is one in which a worker is injured by putting too much strain on some part of the body, or his body is used improperly to complete a task. Typical overexertion injuries are caused by: a) manual handling of objects. The weight of the object may not be an issue; the injury could be the result of the repetitions associated with the job; b) using extreme force on an object that may be stuck or frozen in place; c) using an unsafe position while completing a task; and d) attempting to support or control product or equipment that is off-balance or is falling. For those readers who have a desire to use job hazard analysis for ergonomic improvements, the overexertion type of injury is related to this problem. For repetitive motion claims, the use of the term "injury" in regard to carpal tunnel syndrome is incorrect. Carpal tunnel is considered an illness. Carpal tunnel involves repetitive strain and overexertion, as with most ergonomic issues in the workplace.

Figure 6.27 Forklift operator standing on elevated platform
This forklift operator is not wearing his safety harness and does not have the side arms down on the picking platform. If he were to fall, the consequences could be tragic.

Examples of overexertion injuries are:

• An auto mechanic is preparing to check the brakes on a large truck. Rather than loosening the lug nuts and lowering the vehicle to remove the large tires near the floor of the garage, he took it upon himself to "manhandle' this very heavy tire from the truck and lowered it to the floor. He suffered a herniated disc because of the heavy and awkward lift.

• Two workers are delivering a refrigerator to a second floor apartment. In the process of handling the dolly and refrigerator, one worker injures his lower back.

- An assembly line worker suffers carpal tunnel syndrome as a result of placing plastic protective sleeves on books. The job involves a bonus and the worker admits that she has had difficulty with her wrists while working at that job for past several months.

- A punch press operator complains of sore shoulders as a result of reaching up to activate the control buttons. A consulting safety engineer evaluates the machine and informs management that they will have to place the two-hand controls at waist level to eliminate the hazard. The controls are moved and the problem is resolved. In addition to the worker experiencing less fatigue as a result of the relocation of trip controls, the productivity of the machine also improved.

- A machine operator is required to use a pair of pliers to place and remove parts in a punch press. The pliers hold the part and the worker has to grip the pliers to place and remove each part. The constant opening and closing of the pliers has caused wrist and forearm pain. The pliers are replaced with a magnetic tool that is equipped with a trigger release and the problem is resolved.

- The worker in Figure 6.28 is straining to pull down a poorly maintained roll-up door.

Figure 6.28 Pulling down on cable for roll up door

This worker is struggling to pull down a stubborn roll-up door.

There are some basic improvements that can be used to minimize or eliminate overexertion injuries.

1. Instruct workers in the principles of ergonomics. Develop a set of written guidelines to allow supervisors and workers to correct problems with lifting, grasping, reaching, holding, moving, pushing, and pulling problems.

Figure 6.29 Incorrect lifting technique

This photo is a deliberate example of the incorrect way to lift.

2. Use correct lifting techniques at all times. Using the incorrect posture can cause a back injury regardless of the weight of the object. Safe lifting involves coordinated lifting of loads as well as the use of a helper when necessary. Figure 6.29 identifies a poor lifting technique; all the weight is directed toward the lower back. Figure 6.30

identifies the correct method of lifting: back straight, allowing the legs and arms to do the lifting.

3. Use mechanical devices to lift, move, and position loads. In the long haul, it is less expensive to install machines that lift or tilt than to pay for the costs of workers' compensation and lost productivity.

4. Evaluate the tasks performed at each machine and select ergonomically correct hand tools. The correct tool can easily grip or hold a part so that a worker does not have to reach into the point of operation.

Figure 6.30 Correct lifting technique
This worker is demonstrating a correct way to lift.

5. Require the use of PPE to prevent injuries caused by product handling.

6. Where possible, mark the weight on containers to prevent problems associated with lifting and handling. For loads being moved by hoists or cranes, identify the center of the load.

7. Design workstations involving video display terminals to meet the guidelines recommended by NIOSH.

8. Train workers to use mechanical methods to remove parts that are stuck.

Figure 6.31 Reaching into a picking bin
When possible, lower stock to the waist to shoulder area to make picking and selection easier.

9. Place parts in storage areas that can easily be accessed. Workers should not have to strain to reach parts. They should use ladders where possible, and should lower the picking levels to the waist to shoulder height when possible. In Figure 6.31, the worker has to reach high into the pipe bin to retrieve product. If the bin was lowered, the picking would be more efficient.

10. place handles on tote boxes, baskets, or portable bins for easier handling. Limit the weight that each of these containers can hold if a worker has to move and carry them.

Environmental Exposures

Environmental exposures involve radiation, fumes, gases, vapors, mists, dusts, temperature extremes, oxygen deficiencies, and noise. A worker can easily suffer an illness or injury as a result of exposure to any of these items. Exposures can come from welding fumes, carbon monoxide gas, paint thinners, asbestos dusts, cold storage warehouses, and x-ray machines. The results of an environmental exposure can cause immediate harm, (acute) such as sulfuric acid, or, long-term harm (chronic) from asbestos fibers. Over the past 30-

40 years, much has been done to measure and identify health hazards in the workplace that are harmful. The use of PPE, ventilation, substitute chemicals, and refined work methods have led to a reduction in workplace illnesses.

It is estimated that approximately 60,000 workers die each year from health-related diseases. Many of the diseases are a part of past exposures to dusts, fibers, and chemicals. In many cases, workers did not realize that they were working with chemicals or processes that would cause diseases when they reached retirement age. The effects of the earlier exposures manifest themselves at a later time in their lives—perhaps even 15 to 40 years after the exposure. Industry now has sophisticated methods and machines for testing the air in the workplace. Much more can be done to improve the environmental conditions in factories and construction sites.

Examples of Environmental Exposures Are:

- A welder has been assigned to a project of wire welding galvanized sections of pipe. There is no direct exhaust ventilation in the room. He completes the day of welding and while at home develops chills and flu-like symptoms. When he returns to work the next day he is told by a co worker that he developed metal fume fever from the welding on the galvanized pipe.

- A forklift operator driving a propane-powered machine develops a severe headache and a feeling of nausea after lunch. It is discovered that his forklift was emitting high levels of carbon monoxide and he had been affected by the gas as he loaded and unloaded trucks.

- A painter develops a serious skin rash on his hands after several weeks on the job. When his work procedures are evaluated, it is discovered that he is dipping his hands into the paint thinner to clean them. His actions are allowing all of the natural oils on his skin to be removed. He is also failing to properly wash his hands during the work shift.

- A punch press operator has his hearing checked with an audiogram. He has not been wearing his earplugs properly and the results of the audiogram show that he has had a significant hearing shift. He is advised of the problem and re-instructed on how to wear his hearing protection.

- An auto mechanic is using a degreaser on parts being readied for the scheduled repair. A safety committee representative reads the material safety data sheet and discovers that a safer degreaser is needed. The chemical is changed which provides for more protection for the mechanic.

- When a car is being serviced for the inspection or replacement of brake shoes, a wet wash-recycle method is required to capture the brake dust. The technician in Figure 6.32 can be exposed to microscopic particles of brake dust if the brake drums are not washed out before removing them.

Figure 6.32 Worker exposed to brake dust
This automotive repair technician can easily be exposed to brake dust from servicing brake shoes. He should also be wearing safety glasses.

There is much that can be done to provide a safer workplace by improving environmental conditions.

1. Utilize certified industrial hygienists or other safety and health professionals to measure the atmosphere of the workplace. Inform the workers in regard of the sampling results. Conduct routine testing for air contaminants.

2. Substitute chemicals with higher flash points and high TLVs for workplace chemicals that are more harmful. Always use safer substitutes.

3. Maintain a material safety data sheet on each chemical in the facility. Maintain a complete listing of all the chemicals and place an appropriate tab with a letter or number on each sheet.

4. Provide the appropriate PPE for the protection for all workers. Train all employees to properly care for and wear the protective clothing.

5. Provide the appropriate ventilation at all workstations that require the evacuation of air contaminants. Design equipment and workstations to maximize the capture or evacuation of air contaminants.

6. Isolate and separate chemicals to prevent reactions that could explode or cause any type of incident. Store chemicals and evaluate them per manufacturer's guidelines.

7. Maintain eyewash and whole body showers to provide safety measures in the event of a chemical splash incident.

8. Provide signs and warnings to instruct and guide workers.

9. Tune up powered equipment and add catalytic converters to reduce carbon monoxide gas.

10. Measure the atmosphere in confined spaces. Follow OSHA's confined space standard as a guide.

11. Take precautions to contain chemicals in their containers. Require that lids and enclosures be kept in place to prevent vapors from escaping.

12. Provide spill cleanup kits where chemicals are stored and handled. Train workers to react promptly to a spill and the precautions and safeguards involved in clean up; especially the use of PPE and disposal procedures.

13. Require that all workers report each chemical-related incident so the circumstances can be corrected. If workers feel any symptoms of contaminant exposure, require them to report it immediately.

14. Maintain procedures for opening containers and equipment to prevent exposure. Ensure that workers are wearing the proper PPE.

15. To reduce noise exposure, install baffles and curtains around noise producing equipment or processes. Place compressors and other similar machines away from production areas. Consult with professionals regarding measures to reduce or control noise.

16. Any processes that involve radioactive hazards call for strict controls such as shielding, means of detection, and protective clothing.

Summary

The means of how to identify injury causes, the types of injuries and illnesses caused by these incidents, and practical ideas for making the workplace a safer place are included in this chapter. By breaking down the source of injuries into these 11 categories, supervisors know exactly what type of hazards to look for in each job. Applying the principals identified in this chapter will go a long way towards creating safer worksites, and safer workers and supervisors. Once the means of identifying these contributors are understood, there should be fewer injuries because of the improvements that take place.

Chapter 7

Injury Prevention and Investigation

The main purpose of a job hazard analysis is to prevent injury or illness. In many ways the JHA can be seen as a preemptive injury investigation because it attempts to identify causes of injuries and illnesses before they occur. Because early detection and prevention is important, preceding chapters looked closely at the causes of injuries and preventive measures to be taken to avoid them. This chapter will focus on the nature of injuries, and the investigative process that is to be followed after an injury occurs.

Two forms that are necessary to injury reporting and investigation have been included as figures in this chapter. Both supervisors and employees should be familiar with the Incident Reporting Form, Figure 7.1. Everyone should be encouraged to complete incident reports when necessary. Those reports that involve maintenance-related items should be forwarded to the maintenance department for corrective action. At the conclusion of an injury investigation, supervisors should complete the Injury Investigation Report, Figure 7.2. Management will discover that increasing the emphasis on correcting and investigating everyday incidents will decrease the frequency of property and product damage, close calls, unsafe conditions, and unsafe behavior.

Management should consider injury prevention as equal to the efforts an organization places on quality, productivity and budgeting. Management must always keep in mind that all injuries have a cause and that the causes must be determined in order to prevent future injuries. Injury investigation and cause analysis are only half of the corrective action. All of the causes have to be corrected, not just identified.

Should management wait for injuries to occur before stepping forward to offer corrective action? After all, management does not have a crystal ball to alert them when the next injury will occur, so they tend to wait for an injury before taking any corrective action. When evaluating any injury prevention program, there must be procedures in place for looking for the cause of an injury before it occurs, and if it does occur, immediately after it has taken place. Much can be done to prevent an injury from occurring in the first place. Also, must the injury be serious or life threatening for management to require an investiga-

SUPERVISOR'S REPORT OF INCIDENT INVESTIGATION

WHAT TYPE OF INCIDENT OCCURRED?

Property Damage   Fire Loss   Product Damage   First Aid Injury   Unsafe Condition Observed

Unsafe Behavior Observed   Near Miss / Close Call   Chemical Spill / Clean Up

FACILITY / LOCATION: _____ DEPARTMENT: _____

DATE OF INCIDENT: _____ NAMES OF INDIVIDUALS INVOLVED: _____

_____

_____

1). DESCRIBE WHAT OCCURRED OR DISCOVERED: _____

_____

_____

_____

2). MACHINE OR EQUIPMENT INVOLVED: _____

3). EXTENT OF DAMAGE / COST OF REPAIR OR REPLACEMENT: _____

_____

_____

4). HOW MUCH TIME WAS LOST IN CLEAN-UP OR DOWNTIME: _____

_____

5). HOW MANY EMPLOYEES WERE INVOLVED IN THE INCIDENT: _____

6). WAS ANYONE INJURED: YES   NO

7). WAS FIRST AID ADMINISTERED: YES   NO

8). WAS THERE A FIRE: YES   NO

9). WAS IT AN ERGONOMICS ISSUE: YES   NO

10). WAS THE EMPLOYEE(S) WEARING THE REQUIRED PPE: YES   NO

11). WHAT STEPS WERE TAKEN BY YOU TO PREVENT RECURRENCE: _____

_____

12). WHAT STEPS HAD TO BE TAKEN BY OTHERS TO PREVENT RECURRENCE: _____

_____

13). WAS A HEALTH HAZARD OR ENVIRONMENTAL EXPOSURE INVOLVED: YES   NO

14). CAN THIS INCIDENT OR EVENT TAKE PLACE AGAIN: YES   NO (Explain on Reverse Side of Form)

15). PROVIDE ADDITIONAL INFORMATION, COMMENTS ON ANY FACTORS ON THE REVERSE SIDE.

16). IF NEEDED, HAVE YOU PROVIDED A DRAWING / SKETCH OR PHOTO: YES   NO

SIGNATURE: _____ DATE COMPLETED: _____

Figure 7.1 Incident reporting form

SUPERVISORS INVESTIGATION REPORT FOR INJURIES AND ILLNESSES

A. INFORMATION ON THE INJURED EMPLOYEE

1 Name (last name first)	Social Security No.	2. Department where injury occurred	3. Age of employee
4. Facility Name	5. Section/Unit	6. What machine or operation was involved	

B. INJURY INFORMATION

7. Date of injury	8. Date reported (day, month, year)	9. Time of injury AM/PM	10. Occupation or job assignment at the time of the injury
11. Type of injury and body part injured	12. Date lost time began	13. Date released to return	14. Restricted work required
	15. Length of company service	16. Length of service in position worked at the time of injury	
	17. Date of death	18. Time of death	

C. DESCRIPTION OF THE INJURY OR ILLNESS

19. What job was employee performing	20. What is the JHA number and title
21. What basic job step was being performed	22. What is the job step number in the JHA

23. Describe what occurred to cause the injury. Include information on: the employees physical position, how he was performing the job, what caused the injury, what did the person make contact with or what ergonomic factors contributed to the injury.

D. CAUSES OF THE INJURY OR ILLNESS

24. What were the causes of the injury or illness. List at least three.

25. Make a check mark in the boxes below to identify those items that contributed to or where responsible for the injury or illness.

THE INJURED PERSON OR PERSONS:
- Did not know of the hazard
- Did not know the safe procedure
- Tried to avoid effort
- Tried to avoid discomfort
- Failed to pre-plan job
- Was emotional
- Was fatigued
- Just returned from vacation
- Was just assigned to the job
- Was recently hired
- Had poor vision or hearing
- Had a physical handicap
- Was ill at the time
- Failed to wear PPE as required
- Failed to ask for assistance
- Suffered the same injury before
- Had insufficient job skill
- Lifted too much, or too heavy
- Overreached
- Failed to inspect work area
- Failed to lift load properly
- Additional underlying causes
- Unable to form an opinion of the underlying causes

Figure 7.2 Injury investigation form

Figure 7-2 Continued

26. What in the employees surroundings / work area contributed to the injury:

⊔ Worn out through normal use	⊔ Unsafe basic design	⊔ Congestion, lack of storage space	⊔ Inadequate illumination
⊔ Abuse or misuse by user(s)	⊔ Unsafe construction	⊔ Exposure to corrosion	⊔ Exposure to vibration, etc.
⊔ Required inspection not carried out	⊔ Required clean up not carried out	⊔ Weather conditions, natural causes	⊔ Exposure to temperature extremes
⊔ No inspection required	⊔ No clean up required	⊔ Inadequate ventilation	⊔ Failure to check before using
⊔ MSDS not reviewed	⊔ Guarding missing	⊔ Awkward positioning	⊔ Other

E. STEPS TO PREVENT RECURRENCE OF INJURY

27. Check all of the actions required to prevent future injuries:

⊔ Revision of JHA necessary	⊔ PPE Required	⊔ Improved design or construction	⊔ Work are to be monitored
⊔ Install lighting	⊔ Complete a new JHA	⊔ Place chemicals in cabinets	⊔ Orientation needed
⊔ Re-instruction of worker	⊔ Formal reprimand	⊔ Verbal warning	⊔ Temporary reassignment
⊔ Install guarding	⊔ Add ventilation	⊔ Remind other workers	⊔ Improved inspection
⊔ Modify inspection forms(s)	⊔ Clean up needed	⊔ Purchase part / item	

28. Describe details of corrective actions:
  (a)
  (b)
  (c)
  (d)

F. IMPROVEMENTS BEYOND YOUR AUTHORITY

29. Is there something beyond your authority that needs to be recommended for improvement:

30. Name the individuals involved in improving the conditions or procedures identified in Section E.

Name of Person(s):	Date to be completed:	Actual completion date:
(a)	(a)	(a)
(b)	(b)	(b)
(c)	(c)	(c)
(d)	(d)	(d)

G. ADDITIONAL INFORMATION

30. Witnesses, phone numbers, weather conditions, etc.

Reported By:	Title:	Reviewed By:

tion? Therefore, management must be willing to correct all hazards before they cause an injury. Sometimes, the simplest elements that are looked upon as unimportant can be the cause of a serious injury.

What Are Injuries?

An injury is "an unexpected occurrence, which usually involves a physical contact between the worker and some object, substance, or exposure condition in surroundings that interrupts work activity." Some may question the use of the word "unexpected" in the definition. The injury usually occurs unexpectedly during a process with which the worker is familiar. As an example, in an auto repair shop a technician was using a cutting torch to remove a nut from a ball joint when he suffered a serious eye injury. The safety rule in the shop required that cutting goggles be worn when any torch work was being performed. He knew it would only take a few seconds to heat and cut off the nut. What he did not expect was the hot blob of metal striking him in his eye. He later said that the injury came as a complete surprise because he had performed that procedure, even though it was performed unsafely, hundreds of times. It stands to reason that if the mechanic knew he was going to be struck in the eye with molten metal, he would have worn his cutting goggles. In this example, the contact was from the hot metal and it interrupted work. Not only was a good mechanic out of work for several months, but his talents were missed by his coworkers and the customers he served.

Injuries are almost always unexpected by the persons to whom they occur, whether the injury occurs in the workplace, at home, or on the highway. If the injuries were expected they would most likely not happen. The worker that expects an injury to occur will do something to avoid it. If a worker expects to be struck in the eyes by sparks while using a bench grinder, he will take steps to prevent it from occurring. A full-face shield, goggles, or safety glasses are needed in this example for protection. The spark shields that are required at the top of a bench grinder should not be relied upon to provide all the protection that the worker needs.

Injuries involve some form of work interruption or production loss. Depending on the nature of the injury, the loss could be calculated in minutes, hours, days, or months. A disabling injury to a key worker, such as a machinist that receives high wages, could involve several months of loss to the organization. There are other associated losses from workers leaving their machines to seek help or to provide first aid. Management should realize that these production losses would not occur if the injury did not occur. Injuries and incidents are symptoms of inefficient methods, tools, equipment, materials, machines, or work areas. There are supervisors that do not regard injury prevention as a part of their job; their belief is that their job is to get the production out the door. They fail to realize that production, safety, and quality go hand in hand—what improves safety also improves production.

Not all injuries are alike; that is, the degree of injury or disability will vary as a result of many factors. Injuries or incidents can be first-aid related, or they can be incidents that cause product and property damage. Employees can receive an injury that will require a one-time trip to a doctor, which could result in lost time or even a fatality. Incidents that

occur repeatedly in the workplace every workday should receive as much or more attention as injuries that require medical care. There are contributing factors to all of these
injuries, and it is up to management to recognize them and correct them.

Contributing Factors to Injuries

There are many contributing factors to workplace injuries. The short list below is only a
portion of all of the considerations that should be made from existing information that
should be put to use as a means of prevention. If management takes the time to check the
information and contributing conditions, steps can be taken to prevent needless losses.
The individual items in the list below should be discussed with supervisors to provide
guidance for prevention. Business conducts prevention analyses every day. They discuss
various costs associated with products, processes, or projects, and steps are taken to
prevent future economic loss. Why not use this same thinking to prevent the next injury?

Contributing Factors

- Specific machines
- Night shift versus day shift
- A particular employee
- A particular department
- A particular powered industrial truck
- Lack of employee training
- Lack of supervisory awareness
- Housekeeping
- Lack of maintenance
- Lighting
- A lack of discipline
- An abundance of first-aid injuries
- Extensive product damage
- Extensive property damage
- Lack of accountability regarding investigative protocol
- Lack of personal protective equipment
- Ergonomic issues
- Chemicals or other environmental factors

Specific Machines

Injuries from machines can be devastating. Even though machine-related injuries do not
occur with the same regularity as, say, a strain or sprain type injury, these injuries are
serious when they occur. It is important to check past records to determine if the same

machine or a similar machine in another department or facility has produced past injuries. There may be similar machines in other departments or facilities which may need the same attention to prevent a similar loss. Only a search of the past records will reveal this important information. Within the injury investigation, set aside a portion of the investigative process to check past losses from insurance records and workers' compensation payouts. If the hazards on a particular machine are not corrected / eliminated, there will most likely be another similar injury.

At a metal fabricating plant, a serious injury occurred at a machine used for rolling aluminum sheets. A worker caught his hand in the unguarded rolls and lost several months of work from the injury. The investigation by a safety consultant from an insurance company revealed that this was an old but necessary machine that had never contained guarding. Supervisors, working with the safety committee, designed a guard that prevents anyone from placing their hands in the rollers. Figure 7.3 portrays the guarded machine.

Night Shift Versus the Day Shift

Injuries occurring throughout the working day deserve to be analyzed to determine the times of the day and the shift of the work being done. This information should be recorded and evaluated. Does the second shift have more injuries than the first shift even though the first shift has 50% more workers? Is it the supervision? Are supervisors trained in loss prevention and injury investigation? If an investigation asks these types of questions the true causes of the injury may be uncovered. Since there are usually more management staff present during the day shift, injuries are less prevalent during this shift.

Figure 7.3 Guarding on sheet rolling machine
This aluminum rolling machine is safely guarded.

A Particular Employee

There are some employees that are repeatedly injured. In other words, they have a history of having injuries each year; and in some cases, several injuries each year. Some would call these individuals "accident prone" workers. However, most of us are accident prone if we have our thoughts focused on personal problems such as a sick child, a late mortgage payment, or an emotional crisis. With our thoughts on a personal crisis it is less likely that our attention will be on safely driving our car or properly operating a punch press. When our thoughts are elsewhere we are more prone to error and injury. During an investigation the person who conducts the investigation should determine whether or not the worker has a history of multiple injuries. If they are "repeaters," then additional training is necessary. An employer should never make an example of an employee and make them feel as though they are problem workers. It is important to remind repeaters that other employees are doing the same work on a job and never have an injury. Why do they have an injury doing the same job? Also, ask what management can do to make the worksite safer for the worker. Never make a worker feel as though management has just put a dunce cap on their head because they were injured.

A Particular Department

Each department in a plant or at a worksite is faced with different hazards, work processes, numbers of employees, and injury totals. As a result of these variables, the investigation of any injury must consider the contributing causes such as workflow, parts produced, and the movement of workers. Do not complete an investigation without considering all of the exposures and work process movement in a department. Never record the obvious facts without looking beyond the supporting data from the department.

A Particular Powered Industrial Truck

Powered industrial trucks are not machines that are fixed in place. Forklifts move around the plant, and can cause injuries and property damage. If a forklift or other powered industrial truck is involved in an incident, evaluate the inspection records to determine if the vehicle has been maintained properly. Every vehicle has the potential of causing an incident, but some of them have a history of having problems with controls, brakes, steering, leaking fluid, and faulty gauges. Be sure to look beyond the injury caused by the vehicle. Do not omit the unseen details.

Lack of Employee Training

The lack of employee training can be the cause of many workplace injuries. It is not uncommon for an injured employee to comment that he / she did not know of a hazard. In order to safely and efficiently operate a machine, or any other process, training is necessary. Should an injury occur, be sure to inquire as to how much training the worker had on the particular process. A lack of training is a definite contributor to injuries. Do not forget to inquire about the length of training that was programmed.

Lack of Supervisory Awareness

In most cases, the person conducting the investigation of an injury to a worker is the employee's immediate supervisor. If the supervisor has not been trained in the techniques of investigation, many facts and contributing causes will not be recorded. The investigation process is not a science, but it does take knowledge to use detective-like skills to dig out the facts and contributing causes. Every supervisor must be trained in the techniques that can accomplish the best departmental investigation possible. If management wishes to prevent workplace injuries, they must focus on quality injury investigations through the efforts of the first-line supervisors.

Housekeeping

When fast-paced production schedules become the norm, keeping the workplace clean becomes a challenge. In every injury investigation, it is important to include an evaluation of housekeeping in the area in which the injury occurred. Workers can encounter obstacles that can interfere with the completion of their jobs. As a result, an injury that appears to be caused by an unguarded machine might have been partially caused by the worker stumbling on debris that was near the machine. The worker had to compromise his body position to avoid the debris. Therefore, poor housekeeping practices helped to contribute to the injury.

A warehouse experienced a fire in the area below a dock plate. Most likely someone threw a lighted cigarette on the floor and it fell into under the dock plate. This occurred despite the building being a non-smoking workplace. During the investigation, all remaining dock plate cavities were inspected for housekeeping violations.

Lack of Maintenance

Supervisors must be sure to evaluate maintenance procedures when investigating an injury or incident because inadequate or missing maintenance procedures can easily contribute to workplace injuries. Product manufacturers always recommend preventive maintenance to prevent a breakdown of the machine or failure of parts. If machines or products are not given the proper attention according to their manufacturer's specifications, injuries can easily result. Failure of wire rope, brakes on mobile powered equipment, interlocks, and electronic machine guards can easily be traced to the failure of working parts. These failures can contribute to an injury.

Lighting

Always consider lighting levels when investigating an injury. OSHA's standard for workplace lighting has a reputation of being weak, and sometimes does not provide for sufficient illumination for productivity and safety. To make matters worse, some employers take a literal interpretation of OSHA's minimal requirements, and as a result allow a less than desirable level of lighting in the workplace. The manufacturers of industrial lighting systems should be called upon to design and install lighting that not only meets OSHA standards but also provides illumination that enhances the worksite.

A Lack of Discipline

If a plant or worksite has a weak safety program, discipline among employees will most likely be poor. Poor discipline usually results in poor production, quality, and safety. Part of this enforcement of safety practices and processes begins when workers are being hired. If management does not use good job selection techniques in the hiring process, workers that do not have the skills and job knowledge will be hired. How well do workers follow plant rules and safety rules? When discipline is lacking there is usually chaos in the workplace. Good discipline allows for better chemistry as a working relationship among workers as well as supervisors.

An Abundance of First-Aid Injuries

A regular review of first aid cases at a worksite can help in preventing injuries. How can this review be beneficial? In many cases, first aid incidents are precursors to injuries. If a particular job or department is having an abundance of eye injuries, splinters in fingers, or employee headaches, management must step in to resolve the problems. Eye injuries could be caused by missing safety glasses; splinters could indicate that gloves are necessary.

Employee headaches could indicate a problem with carbon monoxide. If first aid logs go unused or are not reviewed, steps toward the prevention on injuries will go uncorrected. A regular review of first-aid logs will pay dividends toward the prevention of injuries and incidents.

Extensive Product Damage

If management is determined to achieve improved quality of product, efficiency, and safety, much can be done by focusing on the causes of product damage. Extensive product damage can be the result of packages falling off of conveyors because of mechanisms that are inoperable and fail to stop the product. Powered industrial trucks can easily damage and destroy product. Poorly trained operators or poorly maintained equipment can cause this. Workers could be using the incorrect methods to produce or package the product being manufactured. During any walk through or inspection of a department, management should react to any discovery of product damage. Damaged product cannot be sold and has an effect on the bottom line. Until the investigation process is used to uncover and correct these deficiencies, management will continue to lose valuable product.

Extensive Property Damage

The majority of property damage that occurs in a facility is associated with powered industrial trucks. Forklifts and other vehicles are heavy and carry loads that can also damage the building and product being produced. In many cases the operators are not trained and the safety rules are not being enforced. Pedestrians are also victims of inadequate forklift training and enforcement. Aisles have to be wide enough for safe passage of the lift trucks. Lighting has to be adequate to allow for safe passage of workers and vehicles. OSHA's new powered industrial truck operator training rules are very extensive and require instructions that will not only result in safer operators but will help in preventing property damage. During any investigation involving a forklift, evaluate the training of operators.

Lack of Accountability Regarding Investigation Protocol

Who is responsible for investigating injuries and incidents in the workplace? Who reviews the completed forms and ensures follow-up to determine whether or not corrective action has been taken? Are supervisors properly trained so that effective investigations are completed? Management is responsible for determining who completes the inspections as well as who conducts a review of the process. This is an excellent opportunity to allow for worker participation. When an injury occurs, allow trained employees to assist in the investigation process. If accountability is not established by management, the injury investigations will be hit and miss.

Lack of Personal Protective Equipment

Each job hazard analysis and injury investigation report should include a requirement for personal protective equipment. If specific safety gear is required for a job being performed, it is important that employees wear the required equipment. If an injury occurs, the individual conducting the investigation must ask the following questions: was the prescribed safety equipment being worn, was it being worn correctly, and was its use being enforced?

Ergonomic Issues

Employers that are interested in job efficiency, reduced workers compensation costs, and fewer injuries should improve the conditions that contribute to ergonomic injuries and illnesses. Supervisors should be trained in the principles of ergonomics so they can identify ergonomic hazards when investigating an injury or while completing a job hazard analysis. Without the proper training and knowledge, employees might experience ergonomic injuries that will go uncorrected if the causes are not identified. We are now living in a period in which there is a keen emphasis on ergonomic issues. Supervisors should evaluate work practices that involve lifting, reaching, carrying, stacking, word processing, and body rotation. Make every attempt to fit the job to the worker—not the worker to the job. Improvements can be made in the height of tables, weight of loads being carried, handles being installed on tote boxes, or lighting and automation, to name a few. Many ergonomic hazards can be identified and eliminated while completing a job hazard analysis.

Chemicals and Other Environmental Exposures

It would be difficult to imagine a worksite or plant that did not use chemicals in its operations. There are many thousands of chemicals in the workplace, and the numbers are increasing each year. Most, if not all, of the jobs or tasks that utilize chemicals, or have some form of association with chemicals, deserve to be considered for a job hazard analysis. Chemicals in the workplace can have insidious effects on the worker. Many workers and supervisors are ignorant of the potential harm that could be involved while using chemicals. Some 90 percent of all contaminants enter the body through the lungs, 8 percent enter through the skin, and 2 percent enter through the digestive tract from swallowing. If a worker is wearing rubber gloves in the assumption that he is being protected from a chemical he is using, he may be inhaling enough of the chemical to cause harm. There are other workplace exposures that the supervisor and worker should be able to recognize as problems and take corrective action to resolve the problem. Noise, carbon monoxide, radiation, lasers, and other workplace hazards are sometimes unseen and therefore do not receive the attention they deserve. While conducting any job hazard analysis, always consider environmental factors and take steps to correct them.

Investigating Injuries

No one can predict the exact time an injury will occur, but there are methods that can be employed to alert management as to where the next injury might occur. A crystal ball is not necessary if sound investigative techniques are used.

For the sake of discussion, consider the example of analyzing the history of vehicle collisions that are occurring at a particular intersection. The county or state has data identifying when these collisions occurred. If they study the data they will arrive at some conclusion for improving the situation. Let us say that this intersection has a two-way stop sign. Should a four-way stop sign be installed or a four-way stoplight? It is apparent that if the state or county do nothing, the collisions and losses will continue to occur. In this example, the authorities do not know exactly when the next collision will occur, but the statistics and trends indicate that the collisions will continue to occur at this intersection if left uncorrected.

Now let us transfer this same logic to the workplace. It can be a construction site, warehouse, manufacturing plant, or retail store. The information that was previously gathered on injuries and illnesses can help in preventing the next events from occurring. The first item of importance is that there are far greater events or incidents in the workplace that result in close calls, property damage, damage to product, and minor first-aid type injuries. Many people accept these events and occurrences as a part of the job and learn to live with some or most of them. In every case, however, there is a loss to the company. If we add up the details from these incidents, the price tag would be staggering. Keep in mind that the example is not for non-recordable or disabling injuries, but for non-injury events. Some of the incidents could involve in-house first-aid, but do not involve outside medical costs.

When an injury occurs in a department or a plant that is on a safety record streak, there is a natural reaction to question the need of an investigation and in some cases to deny that the injury occurred. Professional organizations do not let the quest for a safety record stand in the way of quality investigations.

To prevent future injuries from taking place, concentrate on eliminating the day-to-day incidents. Along with a strong job hazard analysis program, the top means of preventing injuries and illnesses in the workplace are through a stringent correcting and reporting program. A distinction should be made in the terminology used to distinguish incidents from accidents. This author prefers to use the term injuries versus accidents. In this explanation the reader can substitute the word accident for injury. Workplace incidents, not injuries from "accidents" can be:

- The discovery of an unsafe condition.
- The observance of unsafe behavior.
- A fire of any size.
- Property damage.
- Product damage.
- Close calls or near misses, (near hits).

Any or all of these events should be the focus of the safety efforts at a worksite each day. These events are occurring constantly, and until they are resolved, they will continue to occur and result in some form of loss. The biggest loss to avoid is the injury or fatality that could occur as a result of not correcting incidents along the way. Here we will provide a description of the six examples of workplace incidents:

1. The discovery of an unsafe condition is a necessity at every construction site and industrial plant. Unsafe conditions are present each hour of each day. Some are very dangerous conditions, such as missing planking on a scaffold. Others are less dangerous conditions, such as a carton protruding into an aisle. Both situations are unsafe conditions, however, the missing plank on the scaffold could be life threatening. It might take a few minutes to install a missing plank on the scaffold and perhaps

only a few seconds to place the box behind the yellow line. In both cases a potential fall hazard exists. If the dangerous situation is not corrected, an injury or fatality could occur. Each time a supervisor ventures out into the workplace, he should look at the production efforts required to get the product out of the door and he should also look for "things" that could cause harm to someone. The supervisor should also correct those identified hazards as soon as possible. It is so easy to become distracted and forget to complete a task that you know is important. Never walk away from an unsafe condition without alerting someone to it existence. In Figure 7.4, a pedestrian could be injured by a nail protruding from a broken pallet board. When hazards such as these are detected, they should be corrected as soon as possible to prevent an incident.

2. The recognition and correction of unsafe behavior is necessary to prevent personal injury to an employee or visitor. Many times unsafe behavior is not easily observed unless the supervisor places a focus on this issue while walking through the department. However, some cases are obvious. For example, the forklift operator that runs a stop sign, narrowly misses a fellow worker, or takes a corner too fast can easily be observed violating these safety rules. Perhaps a safety glasses rule is in effect at the plant and several workers are observed without their glasses. This is an easy violation to spot. When it comes to wearing personal protective

Figure 7.4 Protruding nails in a broken pallet board

Broken pallet boards with nails in the upright position injure the feet at many locations.

equipment, if everyone is required to wear a certain piece of equipment, it is very easy to spot those that are not in compliance. What is more difficult to detect and correct is the employee that fails to use lockout / tagout on a machine or reaches into a potential pinch point rather than using a hand tool. In many cases, unsafe behavior and the injury that follows is usually not observed by other workers. To be successful in preventing an injury, a supervisor would have to be standing right next to the machine or operator to stop him or her from committing the unsafe act.

3. A fire of any size should get the attention of management because of the potential consequences. Fire is destructive, and even the smallest of fires can evolve into an uncontrolled blaze. If a particular operation or work project has a history of producing any size fire, steps should be taken to prevent and eliminate the causes. Consider the example of a small metal stamping manufacturer that used a small painting line for applying a coating to special parts. The only time this process showed signs of a problem was when maintenance was being performed on the paint line. The maintenance was always performed on the third shift and management did not know that small fires were a usual occurrence in the maintenance process. None of the fires were large or destructive, but they always occurred when the wall coating paper was being removed.

The paper was there to protect the walls from paint overspray. The process only called for stripping off the protective paper to save extensive cleanup time. When some of the paper would not come off the wall easily, a cutting torch was used to warm the glue that bonded the paper to the walls. The small and unreported fires were the result of the paper and paint spray catching fire. On one morning, just as the day shift was coming into work, the entire paint booth caught fire. Smoke filled the plant and workers evacuated the plant into the snow and cold. The fire department, police and emergency squad had their hands full just finding the fire as a result of the heavy smoke. The plant was shut down for several days for clean up. The union contract called for compensation in this type of event. The costs to management were enormous. Management, specifically the supervisors on the third shift, failed to realize that all fires, regardless of their size, are to be taken seriously.

This loss could have been easily prevented if the supervisors would have investigated the regularly occurring fires in the paint booth. After the fire, it was discovered that an easily removable, and somewhat less expensive, adhesive paper could have been used to cover the inside of the paint booth. The fact that a cutting torch was being used to heat paper that contained paint spray should have been the first indicator that the process was questionable. This true story reinforces the reason for using an incident-reporting program. Many big losses will not take place if the little losses are investigated and corrected.

4. Property damage is present in almost every workplace. The visitor that is familiar with industrial safety can easily see the effects of property damage during any plant tour. One of the more common causes of damage to any part of a building or property is the forklift. Where there are powered industrial trucks, there is often property damage. Overhead cranes and hoists can also cause property damage from the loads that are being handled. Damage can easily occur to items such as ladders, stairs, handrailings, doorways, conveyors, walls, windows, piping, valves, support beams, shelving, racking, and machines. When this kind of damage is discovered, it is important to first schedule corrective action. It is also important to convince workers to refrain from damaging the building. Also, if done correctly, workers should be encouraged to report property damage without fear of recriminations. Forklifts are responsible for product and property damage if operators are not trained properly and the work rules are not enforced. The forklift operator in Figure 7.5 has been properly trained in that one of the requirements is the use of the seat belt on the forklift.

Figure 7.5 Forklift operator at controls
Powered industrial truck operators must be properly trained for the new OSHA regulations. Be sure to include a section on the prevention of product and property damage.

5. Damage to product can be more costly than property damage, depending on what product is being produced. If a forklift spears an expensive box of electronic parts, the loss could be in the thousands

of dollars. Just as in example 4, the poor driving practices involved in operating any powered industrial truck can be the result of the lack of operator training. During any training program involving safe operating principles for powered industrial trucks, the reduction or elimination of property or product damage must be a part of the training program. However, if the incidents that cause damage to product go uncorrected, they will continue to result in loss and possible injury.

6. We have all been witnesses to close calls. They can occur on the highway, at home, while on vacation, or at work. Close calls can serve as a wake-up call. Usually our behavior is modified if we stop and think about the potential consequences of the incident. When some of these close calls occur at work, the employee can simply ignore the event or take steps to keep the incident from occurring again. Incidents are precursors to injuries. If we do not take steps to correct the incident that caused the close call, eventually this non-injury event can become a calamity. Investigate all close calls that are identified and fix or correct those conditions that were a part of the problem. In most cases, workers will not report close calls. Safety programs that include employee feedback and participation will encourage the reporting of many close-call incidents.

Levels of Investigation

All injuries and incidents, regardless of their severity, should be investigated. Because the severity of the injury may vary, many organizations establish different levels to the investigation process. The following five primary levels of investigation should be established:

1. First-aid cases. A program along with a form should be used to record first-aid cases such as slight burns, splinters, slight lacerations, particles in the eyes, skin rashes, and the like. On a monthly basis, the nurse, safety committee, or departmental supervisor should evaluate the forms and attempt to provide more safety equipment or to change procedures if necessary. Perhaps a job hazard analysis should be reviewed if it can improve the safety process. For the most part, there should be no reason to involve the plant manager in the actual evaluation of the forms. However, it is advisable to provide management with a summary of the findings and trends.

2. Incidents that do not produce injury, but instead result in property loss, damage to product, small fires, the discovery of an unsafe condition or unsafe behavior, or any near miss, deserve to be investigated. The immediate supervisor or employee should investigate specific incidents in a particular department. In many cases the cause of the incident can be corrected without too much trouble. All incidents are warnings that, if not promptly acted upon, can easily become recordable injuries. Today's close-call can be tomorrow's fatality if steps are not taken to ensure that the hazard has been corrected. A basic, simple form (See Figure 7.1) can be developed and used to document the investigation of incidents. These completed forms can also be used in safety meetings to remind workers at a particular machine or operation of the hazard. Photos can also be taken of various damaged machines, etc., to document the incident.

3. Non-serious injuries must be investigated, but the process should not require a team of investigators. As an example, let us use the case of a worker that has something in

his / her eye. The worker usually has to go to a clinic for treatment. In many cases, the injury is the result of the worker not wearing safety glasses, goggles, or a face shield. It is possible that the wind blew something into the worker's eye even though he was wearing his safety glasses. Injuries such as these—that only involve a trip to a clinic and a return to the job without any lost time—do not provide many details. The information involved in the claim does not include many details either. Surprisingly, there are many such claims in industry. It is advised to use the safety committee to assist in the investigation. When the workers help to analyze the details, a good investigation should be the result of this effort.

4. At this fourth level, injuries requiring a very focused investigation are usually serious and involve extensive lost time. They can include fractures, amputations, burns, contusions, illnesses, sprains, or strains. Injuries and illnesses at this level usually cost the company a large amount for medical care and indemnity payments to the injured worker. As a result of the serious and expensive nature of the claims, a team effort should be used to investigate and record all of the pertinent data. It is recommended that the facility manager lead the investigating team. The immediate supervisor of the injured worker and a member of the safety committee should all be involved in the process. Injuries that are serious involve more investigation time and effort to truly arrive at the best information possible. As with all investigations, the causes are to be corrected once they are discovered.

5. Level five involves investigations of the most serious of injuries, illnesses, or fatalities. The only redeeming factor at this level is that this type of loss, for the most part, does not occur very often in the typical facility. This type of claim is by far the most costly. When a disaster such as this occurs, not only does local management become involved but outsiders will also share in the investigation. OSHA would have to be called within eight hours. State and local officials would also become involved. The investigation process will not only include more individuals, but the written material will be more extensive than usual. Newspaper headlines and adverse publicity could impact the company. There would also be substantial indirect costs involved. Very few organizations plan for such a test of their program. Management must make every attempt to prevent such a disaster from occurring in the first place.

Sources of Information

An injury or incident investigation is a systematic effort to establish all relevant facts and necessary information related to the "who, how, when, where, and the why" details associated with the injury that occurred. The trained supervisor should know in advance what questions to ask and the details he / she is looking for. The trained supervisor will gather the facts through the following sources:

*   The injured worker
*   Any witnesses to the incident
*   The scene where the injury took place
*   Any vehicles, machines or processes that were involved
*   The adjacent work area

The skilled investigator will use a variety of sources to investigate and arrive at a conclusion that logically presents the facts and provides information to keep the next injury from occurring. The supervisor should not accept everything that is told to him at face value but determine if the explanations match up with the type of injury, the working conditions, and the items that are observed, many of which are identified above. The supervisor's personal experience, coupled with the information from witnesses and the injured worker, should result in a professional investigation. At this point the supervisor records all of the facts and his findings on a specific form designed for this purpose. The supervisor should be more than a collector and a reporter of the information provided. The who, how, why, when, where, and all additional data is needed for the completion of the form. An important section of the form will ask, "What can be done to prevent a similar injury from occurring again?"

Depending on the level of instruction provided to each supervisor, they will conduct investigations per management requirements. The process has to be more than just completing a form "because the boss wants me to do it." Some may think that simply reporting an injury to the state workers' compensation system is all that is needed. Each state requires employers to file an "employer's first report of injury" to initiate the processing of a claim. To some employers, completion of this state form represents their injury investigation efforts. These state forms are basic information documents and simply provide a means of documenting the claim, not solving the problem.

Once an injury has occurred, it is important to gather as much information as possible. Make no mistake about it; it is difficult to conduct a quality injury investigation. It takes skills to properly interview injured workers and witnesses, to investigate the scene of the injury, and to put all this information into a well-crafted investigation report. Injuries should be investigated as soon as possible after they occur or as soon as management is notified of the event. In some cases, the incident occurs and the injury is not reported for a day or two. By then the conditions that contributed to the injury would have changed; this is not an uncommon situation. People forget injury details quickly and it is possible for individuals to furnish inaccurate information because of a lapse in time. The injured employee may not be available to give a statement because of being in the hospital or under medication. It is important to recognize this eventuality.

Never allow supervisors to place blame on a worker during the investigation process. The main purpose of any investigation is to ask the appropriate questions, interview everyone associated with the incident, observe all of the physical conditions, and develop a professional summary of all the findings. Preventing recurrence is one of the main objectives of the investigation process. The search for facts is essentially the search for a means of preventing the same injury from happening again.

Interviewing Injured Workers and Witnesses

The interview with the injured worker is sometimes called the foundation of the investigation. The immediate supervisor and the injured worker must collaborate on the facts and events associated with the loss. Everyone would like to have the worker tell his version as truthfully as possible. However, the worker may be reluctant to provide exact information because he may fear being reprimanded, he might be worried about the impact on his coworkers' future workers' compensation, or he may be angry because he was injured as a result of being at a particular job. It is up to the supervisor to remove these fears and let the worker know that he is only interested in gathering facts on the injury and attempting to prevent the next injury from occurring.

There are a few key points in the interview process that readers may wish to utilize for their investigations.

1. Before starting, put the employee at ease and remind him of the purpose of the investigation. As stated earlier, an organization may wish to use one or more key employees in the investigation. Tell the worker that the intent is not to ridicule him or make him look bad. You are after the facts, so ask the employee to freely give his version of what happened. For the most part, supervisors know the people in their department. The sincere supervisor will remind the worker that he would appreciate his cooperation. The employee can easily recognize the tone of the supervisor and can understand the personalization of the meeting. At this point, the supervisor should be prepared to listen and take notes a little later.

2. Allow the worker to give his version of what happened without interrupting him. Ask him what he was doing and how he was doing it. If at all possible, return to the scene of the injury and allow the worker to describe the actions and conditions at that time. If possible, do not turn on a machine if it can cause an injury or distract from the investigation. Wait for the worker to finish speaking before asking specific questions. A reminder to be neutral in your appearance; refrain from grimacing or using body language. Never make insulting remarks.

3. Once the worker has given his version, ask him specific questions. As an example, ask him to show you where he was on the ladder when he fell. Do not coach him by saying things like, "you were on this last rung, weren't you?" The worker's comments may have omitted some key information from the facts of the injury. The only way to gather the facts is to ask key questions without leading or insulting the worker.

4. Repeat the worker's comments back to him from the point that you are understanding them. If some key points appear to contradict others, politely repeat the story back to him so he can respond. Do not try to box the worker in a corner because his story is a little fuzzy here and there. Slowly describe what you understand from his comments.

5. Ask the worker how this same injury can be prevented from occurring again. Remind him that there are similar operations at other plants or on other shifts and those workers are subject to injury if corrections aren't made.

There are two situations in which the injured worker should not be questioned:

1. Do not question an injured worker if it means delaying medical treatment. A supervisor's first responsibility is to seek prompt medical care for the injured worker. If the worker has left the building for medical treatment, the supervisor should survey the scene of the injury and gather as much information from any witnesses as soon as possible

2. Do not question the worker if they are in pain, visibly upset, or emotionally disturbed. The employee may have suffered his first injury and is reacting to the incident. Some people have a low threshold for pain and the employee is reacting to that.

If any witnesses are a part of any investigation process, they will have to be interviewed as well. A witness is someone that saw the injury or someone that knows something about the circumstances, even though they may not have been an eyewitness. Witnesses should be interviewed in the same way that the injured worker is interviewed. Use the same courtesies for a witness that would be used on anyone else. Do not delay interviewing any witnesses. For the most part, interview them individually to gather the best information.

Remind a witness that he / she is assisting in the investigation process and their comments will not result in any disciplinary action against them or the injured worker. If possible interview them at the scene of the injury. Many times witnesses provide fill-in information that was missed in the initial investigation. Double-check their statements to make sure that you understand their comments and that the image of what happened seems logical.

The supervisor should then proceed to complete the written comments on the investigation report. Up to now he has refrained from taking notes because most workers are uneasy when they are talking and someone is writing everything down. This is human nature and this should not handicap the supervisor.

While interviewing any personnel that were either involved with or witness to an injury occurrence is important in the injury investigation, do not neglect the other aspects of the investigation. Do not allow anyone to disturb the scene of the injury. Information can be gathered from the details of the work area. Check lighting, guarding, conditions of the floor, ergonomic indicators, tools, etc. It is important to conduct this part of the investigation before conditions change. A 35mm camera, video taping, or a good detailed sketch can be very helpful when asking questions at a later time or when formulating a conclusion.

Reviewing the Scene of the Injury

If necessary, re-enact the injury as it may have taken place. Above all, never allow anyone to repeat an unsafe action or put any part of his or her body in harm's way. There have been cases in which the injured employee showed his supervisor how the punch press cycled and how the machine amputated his finger. He then repeated the action only to amputate another finger. An employee can demonstrate what they were doing without actually performing the task.

There is a remote chance that the injury scene would have to be reconstructed. All of the clues related to the loss would be placed in the exact places that they were in at the time of the injury. This procedure may be necessary for a level 4 or 5 injury. Once again, never expose any worker to a risk for the sake of gathering information.

Reporting of Results

If the results of the investigation reveal that specific unsafe conditions that contributed to the injury were allowed by the supervisor to be present, he may be reluctant to report this. It is natural for a supervisor to not admit fault or guilt and elect to omit this information from the investigation form. On the other hand, a supervisor may omit certain information because it means that he will have to provide corrective action and follow-up. Perhaps the costs of the claim might be charged against his department. It should be pointed out that this event rarely takes place but management should know of the possibility of it occurring.

Another problem with the reporting process is identifying obvious tools, equipment, or machines as contributing factors. Either the supervisor failed to look closely enough to record all the facts or any of the contributing factors, such as an obvious missing guard, or they are possibly concealing an accepted work practice. If copies of the supervisor's departmental inspections indicate that all guards were on machines, an injury that involved a missing guard should not sit well with the plant manager.

The program must stress that every injury contains multiple causes. If a supervisor fails to identify as many direct and indirect causes as possible, the benefit of the investigation could be lost. There must be a policy that forbids the use of the word "careless." When supervisors use this term they fail to truly recognize the causes of the injury.

There is also a problem with an investigation when the supervisor omits corrective action because he believes that the problem cannot be corrected. The thinking that "if it is not correctable, it is not a cause" must be avoided. There may be serious hazards or environmental conditions that can easily cause more harm, but these hazards will not be discussed and corrected if they are not recorded.

Sample Injury Investigations

Each of the following three examples of common workplace injuries and management's response to them are analyzed to determine what steps should be taken following an incident and what information should be included on an injury investigation report to prevent an injury from occurring again.

1. *The tip of a worker's finger is nipped off at a pinch point on a machine and a guard is added to prevent a recurrence.*

If the machine was ever guarded at the nip point, management should have recognized the missing guard during departmental safety inspections. If no formal safety inspection program existed, one must be developed; not only for the machine in question but for the entire department. If there was a guard, but the maintenance department or an operator failed to install it, there should be a policy to not operate a machine unless all of the guards are in place. Also, was the worker properly trained to identify hazards in their immediate working area? Did management give workers the right to refuse to operate a machine that lacked guarding?

The following should be listed on the injury investigation report to prevent the injury from occurring again:

- Install a guard on the nip point of the machine operation.
- Thoroughly inspect the entire department every 2 to 4 weeks.
- Issue a procedure to the maintenance department to correctly replace or repair all machine guards if they have been removed or if they are defective. During each walk-through of the department, all maintenance personnel are to look at machines to ensure that they are all guarded.
- The supervisor and safety committee representative will conduct an inspection of all machines each month. Each new machine shall be evaluated for safeguarding before parts are produced on it.
- Train all workers and supervisors to recognize hazards in their respective working area.

2. *A forklift strikes several pallets holding electronic parts. The containers fall and strike a pedestrian. The main problem with the forklift was bad brakes, so management had them fixed.*

Approximately 100 workers are killed each year by powered industrial trucks. Another 35,000 workers suffer serious injuries each year. The injuries include amputations, fractures, contusions, and paraplegia. Another 65,000 suffer less serious injuries. Everything should be done to prevent losses from powered industrial trucks.

Each piece of powered equipment such as cranes and hoists, over the road vehicles, and powered industrial trucks must be inspected on a regular basis. Detailed inspections ensure that the essential safety-related features are in working order. OSHA requires that employers require a daily inspection of each powered industrial truck before it is put into use at the start of each shift. If a plant manager would be conscientious about the brakes on his family vehicle, why not be just as cautious with the maintenance of vehicles at work? It is possible that a survey to identify damage to product and property would reveal that thousands of dollars in damage are occurring on a regular basis. Faulty brakes on the forklift was just one of the causes of the injury.

The following should be listed on the injury investigation report to prevent the injury from occurring again:

- Have the brakes corrected on the forklift.
- Ensure that all operators are trained on their specific powered industrial truck and that the training complies with the new OSHA standard 1910.178 (l).
- Initiate a daily documented inspection program of each powered industrial truck.
- Inspect the plant to check if there are any areas in the workplace where loose product could fall on workers if bumped. If found, take corrective action to move the product or guard it to protect it from collision.
- Install convex mirrors in key areas so that pedestrians are warned of vehicle movement.

3. *A maintenance worker "carelessly" traps his finger between a chisel and a defective rotating stone wheel on a bench grinder. He was dressing up a chisel edge while holding it against the grinding stone and the action of the stone pulled his fingers into the point of operation and in between the tool rest. He amputates the tip of his digit finger on his left hand. Management places blame on the worker's carelessness.*

The seemingly innocent bench grinder is usually taken for granted in the workplace. It is a machine that can be found in almost every industry because the bench grinder stone wheel provides an excellent means of taking the rough edge off of metal parts. It is also a machine that is constantly in need of attention. OSHA issues more citations on bench grinder violations than for any other machine.

A bench grinder has several hazardous components and can be a problem if not provided with constant attention. A bench grinder must have two cover guards on both sides of the rotating stones or wire wheels. An adjustable tool rest is required in front of each stone and requires adjusting to 1/8 of an inch of the stone. Upper tongue guards are required at the top of the grinder frame to prevent fragments of the stone from striking the worker in the event of a stone fracture. Upper tongue guards are to be kept adjusted at 1/8 to 1/4 of an inch from the stone. In addition, spark shields must cover the top of the grinder to protect the eyes and skin of the operator. Bench grinders must be affixed to a table and a "ring" test is required on each stone before mounting. The stone is tapped with a tool at 3, 6, 9 and 12 o'clock to ensure that there are no cracks in the stone. If a ringing sound is heard, the stone is intact. A dull thud sound indicates that there is a crack or fracture in the stone.

Management's approach to the serious finger injury did not solve the problem. Calling an employee careless does not provide any meaningful information to the investigation process. The word careless does not offer corrective action. How many injuries that occur to workers are classified as "worker carelessness" on an injury investigation report? When obvious hazardous conditions are ignored and go uncorrected, management invites injuries.

The following should be listed on the injury investigation report to prevent the injury from occurring again:

*   Inspect all bench grinders each day and adjust the parts as required. Being that it takes only a moment to look at a machine, this is not a difficult program.
*   Replace all bench grinders that lack the proper safeguards.
*   Train supervisors, maintenance department workers, and employees to recognize and correct / report problems associated with bench grinders.
*   Improve the lighting at all machines and operations.

Summary

Investigating incidents and injuries requires training and a know-how of specific techniques if the supervisor is going to do a good job. It is strongly recommended that readers consider utilizing a program that places a focus on day-to-day incidents. If the program is conducted properly, injuries should be reduced when a focus is placed on the "seemingly harmless" incidents. Be sure to investigate events that resulted in product and property damage. Many thousands of dollars are lost in each facility yearly because of product and property loss. If enough attention is placed on he small claims, the big claims will be reduced.

Chapter 8

The JHA Program

The JHA program is not complex, but it is involved and requires planning and practice to ensure that the program is working. This chapter provides guidelines for management to follow in coordinating a successful JHA program, including a program checklist. Also provided is a final example of a completed JHA to demonstrate the final written product of the JHA process.

Managing the JHA Program

If a JHA program is to be successful, it must be managed correctly. Management must commit to supporting the JHA program from the beginning and also recognize that it is a long-term process. Management must ensure that safety guidelines, including all of the JHAs that have been completed, are adhered to by all employees to prevent injury and illness. Once started, the job hazard analysis process should go on indefinitely. However, to maintain this momentum, management must review, revise, eliminate, and add JHAs to the program. In addition, supervisors will need guidance on scheduling, completing, and utilizing these documents.

The process of beginning and completing the JHA forms in an organization will depend on the size of the company and the size of the job assignments. To manage a JHA program, management should:

1. Prepare a written program that would include the correct forms for selecting and recording completed JHAs;

2. Ensure that all supervisors and workers have been trained in the JHA process;

3. Include the completion of JHAs in the annual safety objectives or annual salary review process for supervisors;

4. Establish an overseer for the program; someone that has the skill to review and ensure the quality of the program; and

5. Use approved JHAs for training.

Prepare a Written Program That Includes the Correct Forms

The written program includes the complete guidelines on how the JHA program is intended to function as well as the program guidelines. If a new supervisor was just hired and management wanted him to learn the JHA process, this written program should accomplish that goal. The program should be uniform in content and capable of being used at any location.

The written program should contain sections on the following segments of the program:

- An introduction to job hazard analysis and the purpose of JHA. There should be information on the terminology of the program. The use of the word "job" should be discussed.

- An outline of how a JHA is developed.
    a. Select the job to be analyzed from a listing of all the jobs.
    b. Break each job down into steps as the JHA form is being completed.
    c. Identify the hazards associated with each of the basic steps.
    d. Develop solutions for resolving the hazards that are identified. In addition, identify the correct way to do the job.

- Information will be needed on how the supervisor is expected to:
    a. Select the correct job from the master list.
    b. Discuss the development of the JHA with the worker.
    c. Observe the worker for several minutes or machine / job cycles.
    d. Listen to the operator while he / she is discussing how the job is done.
    e. Ask questions once the job has been studied.
    f. Look at the work area to determine if unsafe conditions exist.
    g. Make note of any unsafe behavior that was detected during the observation.
    h. Record the results of the observation. Read them back to the worker for accuracy.
    i. Submit the completed form for review and formal entry on Form # 3.

- Describe what happens to the completed JHA form once it is completed and approved.

- How are JHAs considered in the annual safety objectives or bonus program?

- When does an approved JHA become a part of the ongoing safety program? How, when, and why would it be used.

- Discuss the posting of approved copies.

- Discuss the use of JHAs in the injury investigation process.

- The particular forms that are used for JHA are important in the program.
    a. Form # 1 is the master listing of jobs.
    b. Form # 2 is the rough draft copy that the supervisor would use during an observation.

c. Form # 3 is the completed and approved copy that can be placed at workstations.

- Discuss the need for ongoing efforts in the JHA process.

A written JHA program should be prepared before the program moves forward. As long as it contains the basics, it should be able to get the job done. Once completed, the program is ready to move forward to the training phase. All of the necessary forms needed to maintain the program should be readily available for use.

Ensure That All Supervisors and Workers Have Been Trained in the JHA Process

Safe work procedures are necessary at each work site. However, safety goes beyond written pages in a manual or handbook. Management must ensure that safe work procedures are placed into practice and enforced. Without the correct and complete training, a JHA program will have little effect on making the workplace safer.

The quality of the training for supervisors and employees is paramount. The training program should be developed to use a step-by-step process to walk trainees through the materials. If the program has been designed correctly, the training documents should be comprehensive and accurate enough to do the job. A written plan is necessary so that the training being presented is uniform and complete for each meeting group. Where organizations have multiple shifts or numerous facilities, the program should be presented in a uniform manner. If the initial training needs indicate that it needs revising, make the appropriate changes before additional training is conducted. All of the instructors and students should be reading from the same guidebook. Keep the language simple and readable.

Everyone in the training classes should be provided with a complete document. The instructors should pace everyone through the JHA process so there is an understanding of the program contents, how the program will function, the use of the forms, and what will be expected of each student. It should be pointed out that each student will grasp the JHA information at his or her own pace. Do not expect everyone to automatically understand the process. Each student will have his or her own learning curve.

Overhead slides, are a very good means of providing information on training programs to students. Students can follow along with the instructor, and the printed material on the screen can be viewed and highlighted.

Ensure that all students are familiar with the materials before moving through the training. The classroom provides the basic information on what JHA is all about as well as how to complete a JHA. However, nothing can substitute for hands-on training. Imagine training students to dive off of the edge of a swimming pool. The instructor can talk about how to dive off of the pool edge. He can show videotapes of people diving. The student can also watch others dive off the edge. But, not until the student attempts to dive will he learn what is involved in the process. The supervisor must be prepared to go out onto the factory floor and be able to observe a worker completing a task and record the task as it is being performed. There is a big step between the classroom and the shop floor. It is rather easy to record what is observed and discussed onto the JHA form, but completing one that is correct and free of errors is reserved for those individuals that have developed a grasp of the process. For those that have difficulty with the program, they will need coaching and assistance. Management must continue to recapitulate the need to stick with the basics and review each rough draft of a JHA that is submitted by the supervisor.

Include JHA's in Annual Safety Objectives and Salary Review for Supervisors

Many plant managers or superintendents do not include safety objectives in the bonus program or annual salary review. Progressive organizations usually establish various objectives for members of the staff. Many of the objectives relate to productivity, reduction of scrap, on time production of the product, and meeting sales objectives. Worker safety is usually excluded from the objectives setting process. Management fails to recognize the benefits of establishing objectives to reduce injuries or workers' compensation costs.

How should objectives be selected for the process? To answer this, management should have an idea of what safety issues are important for the facility. Perhaps there is a need for more employee training, improved use of personal protective equipment, a reduction of injuries, or the creation of a safety committee. Objectives are usually tied into wages, and since most people have a desire to earn more wages, there is usually a serious effort on the part of the supervisors to satisfy the objective.

If management decides to include safety as a means of measurement and improvement, careful thought should be placed in how the process will be established. The typical objective that management decides on is a reduction of injuries. If a supervisor had a certain amount of injuries during the prior year, management has a desire to see a reduction in comp claims that are taking place. They then assign a number that represents the target for fewer injuries that are to be achieved. This system is flawed in that a specific number of injuries are expected to take place. If the supervisor exceeds this number, he will not receive his bonus. If he completes the year and has fewer than expected injuries he will receive his bonus. The question to ask is what is the supervisor required to do to prevent the injuries? This entire process could hinge on fate or luck, not good (or poor) safety management in the department.

What is needed as a means of establishing safety objectives is to select key segments of the safety program and expect that the supervisor will succeed in this venture. What are the critical or key programs that are required to prevent injuries? The supervisor's objectives will contain a limited amount of special tools to implement to achieve success. Management may wish to choose from the following objectives for measuring and rating supervisors:

a. Plant or departmental safety inspections / surveys.

b. Personal training for the supervisor such as first aid, CPR, forklift operations, PPE, etc.

c. The attendance at safety meetings.

d. The conducting of safety meetings.

e. The completion of JHAs.

f. The timely and accurate reporting of injuries in the department.

g. The investigation and resolution of incidents in the department.

Based on these potential select elements of the safety and health program, management should be able to choose from the menu of items and select those items that are most important in the program. The next step is to develop a set of objectives and assign a percentage of the total compensation package to safety. The standard percentages that are used usually range between 10 and 20 percent of the entire compensation package. Before management puts their final seal of approval on this program, they should give it one more review for accuracy and function. One of the problems with programs that are incomplete or inaccurate is confusion on the part of the supervisors. Where money is an issue, everyone becomes sensitive to the threat of errors or double standards as to their share of the compensation.

Once the written compensation package has been developed, the next step is to present it to the supervisors or any one else that will be measured on its standards. A good time to initiate such a program is at the beginning of a New Year or fiscal year. Everyone involved should know what he or she is being held accountable for. Those being measured may ask the following questions:

- Will the percentage of compensation for safety be 10, 15, 20 percent or more?

- What items from the safety program will management want each supervisor to focus on during the coming year?

- Will percentages be assigned for full completion of the objectives as well as partial completion?

- Will management still require an objective regarding the reduction in the number of injuries?

- When will the review of the performance of these objectives take place?

- For those that fail to meet minimum standards, will they be disciplined or discharged?

- What if the company is not doing well financially a year from now; does that mean we won't receive compensation for safety improvements?

- Will the poor performance of one or more supervisors affect the compensation of the entire group?

- What if we exceed the established standards?

- What recognition, if any, would the workers receive from the successes from this program?

Select an Overseer for the Program

Once an organization takes on the challenge to utilize the JHA program, they have to realize that the program will not operate on its own. Someone has to monitor the program to ensure that it is functioning properly. Note that, like a ship in the ocean, the program will not function without someone at the helm.

The position of overseer of the program, which is usually added onto other duties in the facility or at the worksite, will require someone that will have the opportunity to evaluate the process to ensure that it is functioning properly. Most companies assign this duty to the safety director. Many safety directors assume that they will be the people monitoring the program. Organizations that do not have safety staff usually assign the task to someone that has a good working knowledge of plant or construction safety.

The overseer has to be able to review each draft of a JHA, make the appropriate comments on the forms, and return them or review them with the supervisors. The quality of the completed forms has to meet a high standard. The jobs have to be checked off of the master list so that the jobs can be tracked.

Use Approved JHAs for Training

Completed and approved JHAs are an ideal tool for training workers and other supervisors. They represent the safe procedures that, if followed, will have a dramatic impact on reducing injuries. If a job or process changes, a new JHA can be used to train employees in the new procedures that are to be followed.

When one or more employees are about to start a job that they are unfamiliar with, they should review the JHA that is associated with the task in order to learn the safe procedures that are to be followed. Some jobs may take place only annually or semiannually. It is easy to forget the safe procedures for a task that is rarely completed.

When training groups on new machinery, have a select group of employees inspect a machine for guarding, noise, ergonomics, etc. prior to allowing them to use the machine for production.

If at all possible, video-tape jobs during observations so that later these can be edited and used for training purposes. If video-taping is not possible, digital photos of the basic steps or important steps in the JHA can be used for training. A few corporations have compiled JHA photo books to use for new employee training. The books contain at least 24 photos so that anyone looking at an illustrated job would be able to understand how the job is performed.

Final Notes for Management:

- Training should begin with a focus on the first-line supervisors. There are some organizations that may use others to complete JHAs, but the typical person completing the JHA is the person that works the closest to the employee.

- Enlisting everyone's help can help to create a master list. Once the supervisors and employees know what distinguishes a job (or task) from an occupation, the number of jobs listed on Form # 1 will grow. In a short time there could be hundreds of jobs listed on the sheet. A recommended method of obtaining entries on Form # 1 is to give copies to supervisors and employees and ask them to add as many job titles as possible. Allow a set period of time, perhaps a week, for the jobs to be listed.

- Once the completed forms are received, the next step is to sort them out by department and enter the jobs on the master sheet. A "master sheet" should be the only source of job listings. Each job should have a number and title. Firms have an option to list jobs according to their severity or hazards involved. The master lists can be placed online to offer everyone the selection that is best for their department. The job listing can be available electronically or hard copy. Where necessary, jobs may be assigned rather than allowing supervisors to make a personal choice.

- A word of caution: tight control over the jobs selected from the master list is important. Several supervisors working on the same JHA can waste time and effort that could be directed elsewhere. In some cases, an organization may wish to have several groups or individuals working on the same JHA. There should be a program rule of control that requires the overseer's approval to take a job from the list to analyze. The list should always be up-to-date, and the individuals working on any JHA should be documented in the event that someone has to be contacted for any reason.

- A set number of completed and approved JHAs are usually required during a certain period of time. As a control on the quality of the completed and approved JHAs, many organizations require one to be approved each quarter for each supervisor. Depending on the number of supervisors in a facility, the completion and approval of four JHAs each year is noteworthy. In time, the number of JHAs in a binder or online becomes an impressive and useful resource.

- Once the JHAs have been completed and approved, they will have to be readily available for use by the supervisors and employees. Place the approved copies online so they can be reviewed at workstations or key locations in the facility. Paper copies should be placed in a 3-ring binder for easy access. Place copies in the binders in numerical order.

- Where necessary, JHAs can be placed near a machine or operation to make the information readily available. If a worker is newly assigned to a job or has just been hired, the JHA can be reviewed before they go to work. A review of the form(s) allows for accuracy. It is possible that a critical piece of information can be left out of safety and operating instructions and an injury or damaged piece of machinery could be the result. Where JHAs are mounted, protect the document and place the JHA on something firm so that they are easily visible. Place those copies that are not mounted in plastic so that they always look acceptable.

The JHA process is useful because:

- If there is a dispute over the issues involving safe procedures, the JHA can be reviewed to resolve the problem. This same principle holds true for a grievance that has been filed by a worker over a production or safety issue. The JHA should serve as a means of resolving the problem.

- If an injury or incident occurs that has to be investigated, the JHA can be of use in correcting the conditions that caused the problem. The injury investigation form should be designed to allow for identifying the JHA and basic steps involved in the injury. Also, the report should require a JHA review by the injured worker, or a new JHA should be written.

- When supervisors are looking for materials for safety meetings or for the safety committee, a review of specific JHAs can be beneficial. The review can involve more than just looking at the document; the group can modify it. JHAs must be periodically reviewed to ensure that they are accurate and current. A review can uncover issues that may have been missed during the initial development of the form.

- If an organization is attempting to become a candidate for OSHA's Voluntary Protection Program, a review of the entire JHA program may be in order. OSHA will want to look at the completed copies in the program. Prior to OSHA arriving to conduct an evaluation, the completed forms should be reviewed. It might prove embarrassing for the VPP team to be asking questions about a machine that may no longer be in the facility.

- JHA is an element within a program. There are other parts of a safety and health program that compliment the process. When all of these various components are used together, the company will enjoy a better safety record.

Longevity of the JHA Program

Those organizations that have maintained a JHA program for many years are usually organizations with good to excellent safety records. The longevity of the JHA process usually results in safer machines, safer workers, a better-educated management, reduced workers' compensation expenditures, and lower injury totals.

Distinguished performance does not come easy. To excel at safety is a never-ending job. Sometimes the program progresses slowly. If there is a management change or the company is spun off, safety performance and achievement can suffer. The progress gained in developing a world-class safety and health program can quickly erode as a result of lowered levels of management support. If there is one program that should be given support and recognition during periods of change, it should be the JHA program. Because the JHA program has no end, it is one of the few programs that will continue to evolve in the industrial setting.

In a factory where machinery is fixed in place and a product is being produced, the program will be modeled for the conditions that presently exist as well as for those that will be present 10 years in the future. Construction projects are different, however. Each day provides different working conditions than the day before. What is recommended for those organizations that are involved in construction work is to complete JHAs for new jobs and save all completed copies for future jobs. At a new jobsite there may be a need to develop new JHAs, but the entire jobsite may not need new copies. Copies from former jobs can be reviewed at safety meetings and modified to fit the current conditions.

Many jobs, such as erecting a scaffold, digging a trench, working off of a roof, connecting steel beams from a height, and using power tools are probably at the majority of sites. It would be difficult to continuously create new JHAs for each worksite. Construction work can last several months to several years. As a result of these time pressures, creating a new JHA for every job at every worksite is difficult.

JHA Program Checklist

The JHA program checklist that follows should be helpful for designing and managing the JHA program. Each of the line items can provide a reminder of a part of the program that needs attention. The descriptions of the elements in the program will be brief. There are sections within the book that will provide more detail where necessary.

JHA Program Checklist

Managing

- ❑ Has management committed to the support and use of job hazard analysis?
- ❑ Has management assigned the job of overseer of the program to anyone?
- ❑ Is this person knowledgeable in injury prevention?
- ❑ Has the overseer been trained to monitor and assure the quality of the program?
- ❑ Has a written program been developed?

Training

- ❑ Can the written program be duplicated and distributed in the classroom?
- ❑ Have the supervisors, or those individuals that will be conducting JHAs, been notified of this new program?
- ❑ Has a classroom or classrooms been set aside for the training program?
- ❑ Is at least one flip-chart and a set of markers available?
- ❑ Is an overhead projector and extra bulb available?
- ❑ Have the employees been informed of this new program and the fact that they will participate in the process?
- ❑ Does the training program include classroom sessions for all workers?
- ❑ Are the JHA forms readily available for use during and immediately following the training?
- ❑ Is the program designed to train workers in different facilities and on other shifts?
- ❑ During the training sessions, could the instructor answer all of the specific questions about the program?

The Job List

- ❑ Was everyone given a time frame to complete their copy of the job list and a firm date set on submissions?
- ❑ Are employees involved in providing ideas for the job list?
- ❑ Has the overseer separated the lists by departments and prepared a master list for each group?
- ❑ Are job titles offered after the final date for list submission and does management accept the job titles?

Conducting the JHA

- ❑ Has the organization decided what method will be used to conduct the JHAs?
- ❑ Are jobs being completed on a priority basis?
- ❑ Are supervisors correctly alerting workers that they would be a part of the development of a new JHA?
- ❑ Are workers assisting in the recording of the JHA?

❏ Is management encouraging the "one-on-one" method of completing a JHA and discouraged the "absentee" method?

❏ Is management taking advantage of situations where they can use the "group discussion" method?

❏ Are supervisors encouraging workers to go through the process of operating their machine several times while they are being observed?

❏ Are workers explaining the job process while working?

❏ Are supervisors asking questions about the job and pointing out any strange conditions or problems?

❏ Are jobs that have been completed at an earlier date reviewed with employees that are about to perform unusual or infrequently performed jobs?

❏ Are ergonomic hazards being recognized?

Correcting Hazards

❏ If supervisors observed any unsafe behavior or conditions during an observation, are they taking steps towards correcting these behaviors or conditions?

❏ Are ergonomic hazards being corrected?

❏ If maintenance work requests have been submitted as a result of an observation, is the work being done?

❏ If any JHAs have been delayed or shelved temporarily until the conditions are corrected, were the JHAs then completed?

Completing JHA Forms

❏ Is there little delay between the initial observation and the submission of the completed JHA rough draft?

❏ Are there six or less words being used in the basic step column?

❏ Are the correct abbreviations being used to identify hazards?

❏ Are supervisors including information on how to safely complete each job step in the third column of the form?

❏ Are supervisors avoiding the common mistakes of providing too many words in the basic step, not identifying the hazards for the job, failing to repeat information from the basic step, and entering hazards into the third column?

❏ Are supervisors being briefed on how to correct their JHAs?

Controlling JHA Forms

❏ Are completed and approved copies being placed on final written JHA forms?

❏ Have completed and approved copies been placed in their appropriate 3-ring binders?

❏ Are approved copies being placed at key operations in the facility?

❏ Are approved copies being electronically filed so that they can be accessed from within the facility or at remote company locations?

❏ When older JHAs are reviewed, are changes being made to update or correct the former copy?

After the JHA

❏ If a safety dispute or grievance is aired, are JHAs used to resolve the issue?

❏ Are JHAs being used for new employee orientation?

❏ Has management noted a reduction in injuries since the JHA program began?

Final JHA Example

With all of the information that has been provided for the correct completion of a JHA, one final review is offered at this time. All falls are to be taken seriously when it comes to prevention. This is why the following job was chosen as a final example of how to complete an analysis. The job is probably familiar to many readers because it is very common in many distribution centers and warehouses. The job being analyzed is, "Safely operating an elevated order picker." This particular type of powered industrial truck allows operators to drive to a specific location in the facility, elevate the platform the operator is standing on, and pick various products for loading onto a pallet.

Probably the most dangerous steps in the job are when the platform has been elevated. Workers have been known to forget that they are on an elevated platform and step off of the platform, suffering from a serious fall. An injury or fatality is sure to be the result of the fall. When the platform is down and the forklift is performing its regular function, it has the same hazards that any other forklift would have.

The following information would be included on the written JHA form for review and approval. While typically the JHA would be written in the three-column format, this JHA will use individual columns as a way to illustrate the process. The photographs in Figures 8.1 to 8.10 depict each of the basic steps that would be seen during a job observation. If readers have a similar machine and model and they wish to develop a JHA, the information provided here would be very useful.

Figure 8.1 Inspecting the forklift
Before the start of each shift; complete a thorough inspection of the piece of powered equipment.

Figure 8.2 Inspecting under the motor cover
Where necessary, remove battery and power covers to inspect for corrosion or any other damage.

Figure 8.3 Inspecting the safety harness
Thoroughly inspect the safety harness for any problems before putting it on.

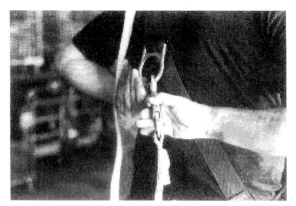

Figure 8.4 Snapping the lanyard to the "o" ring
After inspecting the lanyard and hardware, connect the lanyard snap to the ring on the back of the harness.

Figure 8.5 Lowering the side handles
Be sure to lower the safety arms of the lift truck before starting. Check the lift truck at this time for controls, brakes, etc.

Figure 8.6 Inserting the pallet in the forks and clamp
Drive the lift truck to an empty pallet and prepare to clamp it onto the forks.

Figure 8.7 Clamping the pallet to the forklift
Push on the pallet.locking pedal after the pallet is set in place on the forks.

Figure 8.8 Driving platform forklift to picking station
Safely drive to the various picking stations. Be on the alert for pedestrians and protruding product.

Figure 8.9, Placing product on pallet
Properly lift product from the stored pallets and place them on the forklift pallet. Do not step into the racking or product storage area. Always lower the platform when traveling between picking stations.

Figure 8.10, Driving full pallet to dock area
Once the order has been picked or the pallet is full, Safely drive it to a staging area where it will be shrink wrapped for transport.

Job Hazard Analysis: Safe Use of the Elevated Order Picker

JHA number:

Location:

Person completing JHA:

Person assissting with JHA:

Date completed:

Date approved:

PPE required: Hard hat, steel toe shoes, rubber gloves, leather gloves, safety harness

Basic job steps

1. Complete inspection of forklift
2. Check under the front cover while inspecting
3. Inspect safety harness
4. Hook up lanyard, lower side rails
5. Drive forklift to pallet
6. Lock on pallet, prepare to move
7. Prepare to drive to picking aisle
8. Drive to picking spot, raise platform
9. Pick product and place on pallet
10. Drive full pallet to dock

Potential hazards from job

1. (No hazard)
2. CB Fingers caught between housing cover and frame
3. (No hazard)
4. (No hazard)
5. SB—workers could be struck by the moving forklift

   CW—splinters from pallet
6. (No hazard)
7. (No hazard)
8. SB—workers could be struck by the moving forklift

   SA—operator could strike against racking, objects

   FB—operator could fall to below from platform

   SB—workers could be struck by falling product
9. FB—operator could fall from platform

   O—operator could injure back lifting product

   FB—operator and forklift could tip if overloaded
10. SB—workers could be struck by the forklift

    SB—workers could be struck by falling product

    SA—operator could strike against racking, objects

Safe procedures

1. Use the forklift inspection form and complete the inspection. Check all of the re-quired items on the list.
2. Move motor cover out of the way to inspect the unseen parts of the forklift. Wear rubber gloves if checking the battery. Correctly replace housing. Avoid pinch point when closing it.
3. Inspect the safety harness by using the guidelines provided during operator train-ing. Check the webbing for cuts and abrasions. Check metal parts for damage. If any problems are discovered, report the problem to your supervisor. Put on harness. Adjust for your size. Ensure that all of the parts are in place before moving forklift.
4. Step onto forklift platform and attach the snap of the lanyard on the overhead guard to the ring on the back of the safety harness. Lower the side rails on the platform. Prepare to place forks in pallet.
5. Drive forklift to pallet, drive safely and keep clear of workers and obstacles. Align with a pallet and drive forks into the center of the pallet. If you have to manually handle the pallet, wear your leather gloves. Unsnap your lanyard to handle the pal-let and be sure to reconnect the lanyard to the ring on the back of the harness.
6. Once the pallet has been centered on the forks, step down on the foot pedal at the operator station to lock onto the pallet.

7. Prepare to drive to the first picking location. Select a safe path of travel.

8. While driving to the picking location, be observant of obstacles and other workers. Keep the platform high enough off of the floor for good clearance. Stop forklift directly below the picking spot and raise the platform. Be on the alert for product sticking out of the bins. If protruding product is discovered, put on your gloves and move it away from the travel of the mast and platform. Stay on the platform, do not unsnap your lanyard or remove your harness.

9. While picking product, check it off of the picking list. Never travel to the next picking location while the platform is elevated. Always be alert for workers in the area. Do not allow anyone to be near or under the platform. Do not step off of the platform or crawl into the racking. Do not remove your harness or unsnap the lanyard. Ensure that the weight of the product on the pallet does not cause an overload. Stay off of the pallet. Bend at the knees when lifting product and do not twist your body when carrying a load. Move your feet to rotate your body.

10. When the order has been picked, lower the platform while looking out for protruding product. Observe other workers while driving to dock and follow safe driving rules. Look out for obstacles in the driving lane and to the sides. Set full pallet in place without striking other loads. Return to the picking area and repeat the process.

Comments on the Final Example

This job has high potential for an incident. If an injury were to occur, step # 9 would most likely be the one to cause the injury. Should the worker forget to wear his harness and lanyard or remove it, he can easily fall from the platform. If he steps into the racking or onto the pallet, the risk of a fall increases. If the worker lifts improperly, he can injure his back. Product could be dropped or knocked down. Just lifting the platform to the picking station could result in a collision with any product sticking out of the rack. Steps 8, 9, and 10 could be hazardous if the safety rules are not followed. The other steps contain minimal hazards or no hazard. This is why the job can be deceiving to operators and supervisors. For those facilities that have a similar operation, safety rules for the job should be rigidly enforced.

When completing this example, an assumption had to be made that the pallet was sound and appropriate for the platform. The supervisor performing the job observation should have made sure the pallet was acceptable. It is also assumed that the inspection was the first of the day for that particular forklift, the safety inspection did not reveal any mechanical problems, the amount of product that was to be picked and set on the pallet was well below the capacity of the forklift, and the operator was trained to inspect and wear the harness. These are all items that the supervisor performing the job observation would inspect while in the workplace.

Note that there were several steps that did not contain a hazard, but every step indicated whether a hazard was or was not present. The JHA form should clearly demonstrate that every step was analyzed for potential hazards even if none were found.

Summary

Once the program has started, there will have to be additional management support to keep the program going. Strong management is necessary to ensure that proper training occurs, the job list is kept up-to-date, approved JHAs are controlled, and JHAs are incorporated into company training. The extensive checklist included in this chapter should be of assistance to management.

The example of a completed JHA was provided in this chapter as a final summary of how to identify basic steps, hazards, and correct procedures for completing a job. Note that each of the 10 basic job steps contained six or less words. The hazards column utilized the abbreviations to identify the hazards that are present in the job. In the safe procedures section, a description was offered for the correct method for completing the job.

Appendix A

Safety and Health Programs

Every organization will develop its own version of a safety and health program, which is usually based on a need to reduce injuries or comply with OSHA regulations. OSHA recognizes job hazard analysis as a key element in an effective safety and health program. This appendix offers an overview of the components that are used to develop a successful safety and health program, followed by additional information concerning safety issues and safety performance, including Bureau of Labor statistics that may be of interest to the reader.

Regulations for Safety and Health Programs

Although there is no federal legislation requiring a written safety and health program, OSHA, various state governments, the American Society of Safety Engineers, and the National Safety Council have been advocating written safety and health programs for many years. OSHA, however, does regulate and enforce safety and health standards for millions of workers. Many organizations have in place a safety and health program that includes a written program as a way to uphold safety and health standards. Perhaps in the future a comprehensive program will be required. Currently, however, many employers, especially small employers, do not maintain safety and health programs because they are not legally required to have one.

Federal mandates regarding the development and use of safety and health programs should not be the only compelling force for an employer to implement a safety and health program. Management should seek to protect its single most valuable resources in the building—the employees. Besides the reduction in worker's compensation benefits and the increase in productivity that a safe workplace provides, workers will derive an overall feeling of comfort knowing that they work in a safe environment. So, if an employer carries the belief that he or she is really concerned for each and every worker, it would be advisable to implement a comprehensive safety and health program.

On January 26, 1989, OSHA published "Safety and Health Program Management Guidelines; Issuance of Voluntary Guidelines- 54:3904-3916. The purpose of this document was to provide guidelines to employers to be used in preventing injuries and illnesses. OSHA reasoned that ever since the OSH Act was developed, representatives from OSHA had noted a strong correlation between sound safety and health practices on the part of management, and a low incidence of injuries and illnesses. However, safety and health professionals have estimated that only about 10 percent of businesses have effective safety and health programs. OSHA would like to see this percentage increased to where more employers are using the correct programs to protect workers.

OSHA's voluntary program guidelines encourage the use of job hazard analysis by stating that, "Job hazard analysis is an important tool for more intensive analysis to identify hazards and potential hazards not previously recognized, and to determine protective measures. Through more careful attention to the work processes in a particular job, analysis can recognize new points at which exposure to hazards may occur or at which foreseeable changes in practice or conditions could result in new hazards." Job hazard analysis provides "for the systematic identification, evaluation, and prevention or control of general workplace hazards, specific job hazards, and potential hazards which may arise from foreseeable conditions," which is a key element of an effective safety and health program according to OSHA's voluntary program guidelines.

In addition to the identification of hazards and on-going examination of work processes and conditions, OSHA's voluntary program guidelines emphasize safety and health training that ensures that "all employees understand the hazards to which they may be exposed and how to prevent harm to themselves and others from exposure to these hazards, so that employees accept and follow established safety and health protections."

Recently, OSHA has pursued the development and issuance of a final safety and health standard. At this time it is in draft form only. If approved, this program would become law and employers would be required to establish a written program.

The core elements in OSHA's "Working Draft of a Proposed Safety and Health Program Standard" are:

- Management leadership and employee participation;
- Hazard assessment;
- Hazard prevention and control;
- Worker training, and;
- Evaluation of the program's effectiveness.

Employers would be required to provide systematic control of hazards, and to identify the hazards in new equipment, materials, and processes. OSHA suggests that the analysis be completed before the hazards are brought into the workplace. The complete contents of this proposed standard can be found at www.osha.gov.

Within OSHA's proposed ergonomic standard, there is a definite reference to job safety analysis. Once an employer determines that an injured employees job meets the action trigger, the clock starts for employers to meet the following deadline; note, for the purpose of brevity and the topic associated with this book, the action at 60 days requires a job hazard analysis on the job that was responsible for the injury. Action trigger occurs when a job meets two tests: 1) the employee's injury must meet the definition of a work-related MSD incident, and 2) the employee's job must be one that is plausibly related to the MSD incident and must routinely (on one or more days a week) involve exposure to risk factors at levels described in the Basic Screening Tool. [MSD=musculoskeletal disorder.])

It is apparent that the afore-mentioned safety and health programs and plans drafted by OSHA all have a component for the use of job hazard analysis. The success of a safety program lies in the fact that the injury problem has to be attacked from several different angles. Safety and health programs are not one-dimensional in design.

Safety programs are not created equal in most cases. The culture of the facility and the values that local management places in safety and health programs will shape the program. How successful the program becomes is contingent on management's effort and organization of the program.

The Voluntary Protection Program

In addition to OSHA's desire to have employers initiate a safety and health program, OSHA designed the Voluntary Protection Program, VPP, in 1982. This program was designed to recognize employers that maintained facilities that were showplaces for safety and to provide examples and incentives for other businesses to emulate to achieve this same level of quality. VPP sites have shown that they do in fact maintain safer workplaces. Incidence rates in VPP sites are usually 40 - 60 % lower than the same rates in comparable industry. Some sites have achieved rates that are 80 - 90 % lower than others in their field. At the time of this writing, there are more than 600 VPP sites nationally. When one considers that OSHA is responsible for 6.2 million establishments, the 600 plus VPP sites represent the "cream of the crop."

OSHA realized in the beginning that it had limited resources and could not possibly inspect every worksite. The new federal regulations did not require organizations to develop written safety and health programs. Many safety and health professionals have commented that OSHA would have been better off had they required written programs at their inception. At the present time, legislators are attempting to enact legislation for this purpose. OSHA inspections alone will not ensure a safer workplace.

OSHA has developed very specific requirements for employers that wish to submit a VPP application. An organization should make contact with the VPP coordinator at OSHA's Regional Office and discuss the application process. There may be specific requirements that are part of the local OSHA office that the employer should know about. The OSHA VPP coordinator can answer any questions before the application is prepared.

Once the OSHA office receives an application, it is thoroughly reviewed for content. If any of the required program elements are missing or incomplete, OSHA will contact the work site to discuss the details on how to improve the application. OSHA will not proceed with the processing of an application until all of the necessary information has been provided in the manner that the program requires.

The quality of the materials that an employer has maintained for their safety program should be at the excellent to superior level. Copies of specific program documents, such as safety inspections, audits, emergency programs, hazard communication, lockout/tagout, injury investigation and job hazard analysis must be available, and the program elements should date back for three years. OSHA will want to view these programs as well as their historical use in the overall program. A program that is mandatory and one that comes under a lot of scrutiny from the OSHA team is a contractor's safety program. What should be done to protect each worker when contractors are on their property? The VPP coordinator from OSHA can assist them, or an organization that is a VPP member can help in mentoring.

Employers may wish to prepare the application by using a three-ring binder and placing all of the required documents in tabbed sections so the material is easy to find. Any worksite that is preparing an application for VPP status must prepare a document that provides responses for all of the program components that OSHA has identified. The key components are:

- General Information: Includes site address, facility manager's name, corporate name, union affiliation, if any, number of employees, type of work performed at the site and the SIC code, injury incidence rates, and the lost workday injury case rate.

- Management Commitment and Planning: Requires a copy of the site safety policy, how the safety duties are assigned and structured, who is responsible and accountable for safety, additional safety resources, if any, how contract workers are handled for safety, information on how the employees were notified of VP, and a map of the site.

- Work Site Analysis: How new equipment is evaluated for hazards, facility inspections, facility surveys by professionals, Job hazard analysis, how employees notify management of hazards, injury investigation, and information on the medical program.

- Hazard Prevention and Control: Professional expertise used, safety and health rules, PPE, Emergency planning, and preventive maintenance.

- Safety and Health Training for all employees.

- Employee Involvement: Details on safety committees and how hazards are recognized.

- Program Evaluation: Requires a copy of the prior year's plan for injury reduction, how rates will be reduced, and any additional information the site wishes to provide.

- Statement of Commitment: A statement from the union and a separate statement from management are required.

If the worksite is unionized, the authorized collective bargaining agent(s) must sign a statement to the effect that they either support the VPP application or that they have no objection to the site's participation in VPP. The statement must be on file before OSHA comes onto the site. The statement signed and supported by management is quite descriptive:

"We are committed to doing our best to provide outstanding safety and health protection to our employees through management systems and employee involvement. We are also committed to the STAR program requirements and to the goals and objectives of the Voluntary Protection Programs.

We agree to provide the information listed below for OSHA review on-site. We agree to retain these records until OSHA communicates its decision regarding initial VPP participation."

We will likewise retain comparable records for the period of VPP participation to be covered by each subsequent evaluation until OSHA communicates its decision regarding continued approval."

Numerous discussions with plant safety directors have indicated that a weak or non-existent JHA program is often a stumbling block in becoming a VPP site. The VPP guidelines require that a site have a program in place for at least one year. It should be noted that the VPP process stresses quality and quantity, with a little more focus on quality of materials and program elements. Those organizations that have never used a JHA program are often unable to step right into the process because it is more sophisticated than many other programs. A lack of an effective JHA program has kept some firms from attaining VPP status. They then have difficulty moving forward because of a dearth of training and knowledge.

It is important to keep in mind that the VPP program is voluntary and only those work sites that can qualify for the rigid requirements of the program need apply. These super-safe sites also report lower workers' compensation costs, improved morale, reduced production costs, less product damage and fewer grievances. Information on the VPP process can be accessed from OSHA's web site (www.osha.gov).

Readers may question the benefits of pursuing the VPP program. Why should an employer choose to become a VPP site? Identified below are comments from organizations in the VPP program and from the Voluntary Protection Program Participants Association, VPPPA.

- VPP is an excellent management and business tool. The program contains elements of total quality management methods that recognize excellence.

- Being a member of the VPP establishes a cooperative relationship with OSHA. Business, labor, and OSHA work together to achieve STAR or MERIT quality programs. STAR status is the highest rating OSHA bestows on a facility. MERIT status indicates that there are several parts of the safety and health program that have to be improved upon. However, MERIT status indicates that overall, the organization has an excellent program.

- OSHA and other business organizations work together with the VPP site to gather information on how to develop safety programs and how to initiate best practices.

- When a site becomes a VPP site, they are removed from OSHA's random inspection program for a period of three years. This exemption allows OSHA to focus on those worksites that have poor safety records and programs.

- Being a part of VPP helps to build employee ownership in the safety and health program that fosters continuity.

- Liabilities at the work site are reduced. This would include workers compensation costs as well as reduced third party lawsuits. Being a VPP site is also beneficial to the community.

- When contractors are on a VPP site, the safety requirements are very stringent. One of the specific details requiring admission to VPP status is to have a program that evaluates any non employee group that is on the work site. Safety and health protection is required for both employees and non-employees.

- The annual VPP evaluation helps to quantify the progress of the safety and health program. Management tracks goals and objectives related to the work site. Each VPP site must complete an annual summary of their safety activities and forward the reports to their local VPP representative. Establishing goals and completing recommendations offered by OSHA helps to move an organization forward toward a better program.

- Some additional benefits from the VPP and other key performance indicators include:

  -Fewer grievances,

  -Housekeeping and workplace improvements in image,

  -Overall quality improvements can be measured,

  -Reductions in incidents relating to damaged product and damage to the building,

  -Improvements in program reliability,

  -The costs of compensation claims are reduced,

  -There are noticed improvements in productivity,

  -Worker absenteeism is reduced,

  -The organization assists and serves as a mentor to other organizations that desire improved safety performance,

  -There is an increase in employee pride in their safety and health program; and

  -The organization has an increase for competitive advantage over other organizations.

During the OSHA team visit to the prospective VPP site, employees will be interviewed and questioned regarding various programs that the employer maintains. This part of the OSHA visit causes a lot of anxiety among plant managers. Employees will be interviewed singly or at their machines. Each worker must be familiar with VPP as well as the safety program so that OSHA is satisfied with the quality of the safety and health program. How would your workers respond to the following questions regarding your safety and health program?

A. Background Information

   1. What is your job here?

   2. How long have you been working here?

B. Employee Orientation and Training

   1. Did you receive safety and health training when you began work here? If you did, please describe the program.

      a. How soon after you began to work on your job did you receive training?

      b. How long did the training last?

2. If you did not receive training when you were first hired, or if you were transferred to a new or different job, have you received any basic safety and health training since that time? If you have, please describe the training.

3. Currently, do you receive regular safety and health training?

   a. If so, how often do you receive training?

   b. How long are the training sessions?

4. Are you aware of the company safety rules?

   a. If you are aware of the rules, do they seem to cover everything they should?

   b. What happens if an employee disobeys a company safety rule?

5. What are you supposed to do in an emergency? When is the last time you practiced an emergency drill?

C. Correction of Hazards

1. Do you ever come in contact with any safety hazards?

2. If you do come in contact with safety hazards, please respond to the following questions:

   a. Does management understand the hazards associated with your work?

   b. Has management been quick and responsive to notice hazards and correct them?

D. Reports of Safety and Health Problems

1. Have you ever reported a hazardous condition to your supervisor or other management official in this facility? If yes, please respond to the following questions. If no, we will go on to the next question.

   a. What was the hazardous condition?

   b. Whom did you notify of the condition?

   c. Did you report it orally or did you put it in writing?

   d. Did you get a response? If so, was the response satisfactory?

   e. How long did it take to get a response?

   f. If you did not get a response, did you try again, try someone else or somewhere else? If you had to try somewhere else, please describe what happened.

2. Have you ever filed a safety and health complaint with OSHA? If so, how would you compare OSHA's response with that of the company?

E. The Health Program

1. Do you come in contact with any potentially hazardous chemicals, substances, or harmful physical agents such as noise or radiation? If you do come in contact with any of these, what are they?

   a. Do you feel that management has provided enough protection for you in regard to health hazards?

   b. Is maintenance of release prevention equipment satisfactory? (OSHA would ask this question if the worksite were a high-hazard chemical plant).

2. Have you ever seen industrial hygiene surveying or monitoring being done in your workplace?

   a. Did you see the hygienist only once or is the sampling and monitoring of the work area routine?

   b. If you saw this just once, was it in response to a specific problem? If a specific problem, what was it?

   c. If the monitoring is routine, how often does it take place?

3. Has the company had you examined by a physician? If so, is this done periodically?

   a. If routine, how often?

   b. If not done periodically, what was the reason for the examination?

   c. Did the examination seem thorough?

   d. Did the doctor explain what he/she was doing and why?

   e. If not, did anyone in management explain? If so, who?

   f. Were the results of the examination explained to you?

   g. If so, who explained them?

F. Personal Protective Equipment (PPE)

1. Do you use any PPE such as a hard hat, respirator, gloves, goggles, etc.?

2. Is the equipment available when needed?

3. If PPE is used, is it kept clean and in good repair?

   a. Who is responsible for this?

   b. What protective equipment have you used?

   c. Have you been trained in the use of this equipment? If so, was the training adequate in your opinion?

G. Safety Committees

1. Are you aware of the safety committee (or any other employee participation method) for safety and health?

2. If so, please answer the following questions:

   a. When did you become aware of the committee?

   b. Do you know any of the members? (If yes, please name the members that you know.)

   c. Do you know how the employee members were selected? (If yes, please describe.)

   d. Have you seen them make inspections? If so, does it appear to be thorough in its approach?

   e. What other things do they do?

   f. Would you say this activity is very effective, somewhat effective, or not effective? Why?

H. General Questions

1. Have you ever seen the OSHA # 200 Log of Injuries and Illnesses or a summary of the log? If so, did it seem to agree with your knowledge of accidents and illnesses here?

2. How does this workplace compare to others where you have worked in terms of safety and health? Worse? About the same? Better? Much better?

3. (To be asked in high hazard chemical plants only) Is employee turnover high?

    a. If so, why?

    b. Also, if so, how long does it take a new employee to learn to work safely alone?

4. If your site is approved for this program, OSHA will stop doing routine inspections but will inspect in response to employee complaints, serious accidents or chemical leaks. Under the program, OSHA will come back to evaluate how well things are going as we have done today. How do you feel about that? Do you think it will be okay?

5. Is there anything else you think we should know about the safety and health program here?

Fourteen Elements of a Successful Safety and Health Program

The National Safety Council asked select safety and health professionals their views on the importance of various safety and health practices. The following elements are taken directly from their 239-page book, *14 Elements of a Successful Safety and Health Program* (Chicago, National Safety Council, 1994). For the sake of identification only, the 14 elements have been summarized and abbreviated by using the executive summary portion of each section.

If an employer or a safety committee was looking for a written plan or map to use for developing a safety and health program, the Safety Council's 14 elements will fill that need. Many medium to small employers are looking for ways to improve safety in their worksites. Instead of going through trial and error to create programs, why not use programs like OSHA's Safety and Health Guidelines or the Safety Council's 14 elements? Having this information available will serve as a menu for the components that they require. Every employer may not have the resources to completely adopt these programs in their entirety. The biggest benefit of having a variety of programs to choose from is that the systems have stood the test of time and have proven effective in a variety of industries. If the programs are properly applied to the worksite, there can only be one outcome: improvement.

Element 1: Hazard Recognition, Evaluation, and Control

Executive Summary

A successful safety and health program requires recognizing hazards early and addressing them effectively. Companies can use several tools to effectively recognize, evaluate, and control hazards.

Management must first adopt a proactive approach to safety and health. Senior management's commitment and involvement are crucial factors in the success of a hazard identification, evaluation, and control program. Management has a key role in communicating a safety and health-orientated vision, establishing policies, and allocating resources.

Effective hazard identification requires evaluating all possible hazard sources, including management practices, equipment and materials, the physical work environment and employee attitudes and behavior. Several efficient tools are used in this multifaceted hazard recognition and evaluation approach. They include job safety analysis (JSA), injury/illness/incident investigations, industrial hygiene exposure assessments and systems safety reviews.

After hazards have been identified and prioritized, management must take the next step and implement measures to eliminate or control them. Methods for hazard control include engineering controls and redesign, administrative methods (such as preventive maintenance, housekeeping programs, hearing conservation programs, etc.),and personal protective equipment.

Element 2: Workplace Design and Engineering

Executive Summary

Safety and health hazards are most effectively and economically addressed in the planning and design stage. Building safety and health into the workplace requires the involvement of safety and health professionals and others in planning facilities, processes, materials, and equipment.

Safe workplace design takes into account the optimum physical and psychological compatibility between the employee and the process, methods of operation, equipment, materials, and machinery. A workplace designed and constructed with the employee in mind will have a favorable impact on productivity and quality, as well as safety and health.

A safe and healthful workplace is the result of a continuing process of design, evaluation and modification. Safety and health oversight on workplace design is essential to the elimination or control of hazards.

Workstations with automated processes help to make today's workplace operate more efficiently. But these innovative features may create new safety and health hazards at the contact points between employees and machines. Therefore, these contacts must be evaluated for hazards and addressed.

Element 3: Safety Performance Management

Executive Summary

Even a work environment designed for safety and health will not prevent injuries and illnesses unless people perform their tasks safely. Successful safety management must address performance. Companies committed to safety and health will examine the factors that affect performance and the tools used to measure and improve it.

Successful safety and health performance is possible when the organization has a clear, well-communicated, performance-oriented vision for safety and health. The vision must come from senior management and must be carried out by safety and health professionals, managers and employees, with senior management involvement throughout the process.

The key to high quality and consistent safety and health performance is adhering to a set of established standards and guidelines. These standards can be set from within an organization, imposed by external sources (such as regulatory authorities) or adopted from an external source (such as a standards-setting organization). Holding individuals accountable for adhering to standards emphasizes the importance of and helps to ensure safe performance.

Monitoring performance through performance reviews and appraisals can help management identify problems and solutions. Performance reviews use numerical performance data to evaluate compliance with safety performance standards and to communicate the findings to senior management.

Element 4: Regulatory Compliance Management

Executive Summary

Regulatory compliance is the driving force behind most safety and health programs. Regulatory authorities in the United States and around the world often have differing safety and health requirements.

The regulatory framework for safety and health includes requirements from numerous authorities, including the U. S. government and individual states. Many countries impose safety and health regulations on companies doing business within their borders. The major agencies involved in developing and enforcing U. S. regulations include the Occupational Safety and Health Administration and the Mine Safety and Health Administration.

Effective regulatory compliance requires staying informed about current regulatory requirements and anticipating future issues. Among the information sources for regulatory information are the Federal Register, periodicals, and computer services.

Compliance should be considered only a minimum standard. Successful safety and health programs go beyond compliance and strive to manage risks to worker safety and health.

Element 5: Occupational Health

Executive Summary

The goal of occupational health services is to protect the health and well being of employees. Different companies have different needs, and occupational health programs will vary in complexity from simple first-aid services to a comprehensive set of off-and on-the-job medical and health education services. The following features are characteristic of occupational health programs.

- A successful occupational health program requires the involvement of medical and technical professionals with specialized training in work-related injuries and illnesses.

- All occupational health programs must provide for basic response to occupational illnesses and injuries. Most organizations address this need through a combination of on-site and easily accessible external services. First-aid services should be available when timely medical treatment is not otherwise and to provide prompt treatment for minor injuries.

- Periodic medical examinations are sometimes required to make sure the employee and the job are medically compatible. Where incompatibilities exist, work restrictions will exist and modifications in work procedures, equipment or process may be necessary to accommodate the employee. Medical evaluations are frequently given before new or transferred employees are placed in a given job. Accurate and complete medical records are crucial for monitoring the health and well being of employees.

Element 6: Information Collection

Executive Summary

Information is the key to decision making in all areas of business. For safety and health purposes, data can be used to determine workplace hazards and to identify areas for improvement. Inspections, recordkeeping, injury/illness/incident investigations, industrial hygiene surveys, and performance reviews all generate the information used to evaluate safety and health performance and improvement.

Injuries and illnesses cost organizations directly and indirectly. Quantifying injuries and illnesses in terms of dollars gives senior management a clear picture of their real cost.

Computers and analytical software often can be used to facilitate data collection and analysis. Computers can help manage large databases and general analysis and custom reports quickly and easily.

Element 7: Employee Involvement

Executive Summary

By involving employees in the decision-making process, management affords them a greater sense of "owning" the resulting policies. Ownership translates into increased acceptance of and support for those policies. Management can facilitate a cooperative effort to improve the safety and health program through employee participation.

The team structure that management chooses should fit its particular culture and organization. We discuss the basic concepts that support the successful formation of goal-oriented safety and health teams.

The safety and health committee is a form of employee involvement that is sometimes mandated by law and/or required by organized labor contracts. For this reason, we address committees separately in a discussion of issues common to committee organization, with special attention to union participation.

Management must also foster the involvement of individual workers who do not participate in team structures. Companies can use a number of tools to elicit individual input and cooperation.

Management should validate the value of employee involvement by routinely and formally recognizing those individuals or groups who have made significant contributions to safety and health.

Element 8: Motivation, Behavior, and Attitudes

Executive Summary

Motivation involves moving people to action that supports or achieves desired goals. In occupational safety and health, motivation increases the awareness, interest, and willingness of employees to act in ways that increase their personal safety, that of their co-workers, and that support an organization's stated goals and objectives.

Motivation aims primarily at changing behavior and attitudes, and is generally defined by three factors: direction of behavior, intensity of action, and persistence of effort.

Among the many approaches that companies have used to motivate employees to improve safety and health performance, two general ones have had considerable influence. One, the organization behavior management (OBM) model, is tied directly to the use of reinforcement and feedback to modify behavior. The other, the total quality management (TQM) model, flows from attitude adjustment methods to achieve quality improvement goals in industry.

The ultimate success of a motivational model in changing employee attitudes and behavior depends on visible management leadership. In addition, the motivational techniques used should support the mainline safety and health management system, not take its place. Similarly, evaluation of the worth of these techniques should be measured in terms of how well they achieve their support roles, such as maintaining employees' interest in their own safety, rather than by their effect on injury rates. Three specific motivational techniques are discussed: communications, incentives/awards/recognition, and employee surveys.

Element 9: Training and Orientation

Executive Summary

Today, organizations must comply with a multitude of new regulations, which typically include training requirements and standards. In the united States, OSHA (Occupational Safety and Health Administration), MSHA (Mine Safety and Health Organization), DOT (Department of Transportation), FAA (Federal Aviation Administration), USCG (U. S. Coast Guard) and the EPA (Environmental Protection Agency) specify training standards in their regulations. Similarly, specific directives in the EU (European Union) framework directive include training requirements. ISO 9000, which is a set of universal quality system standards, requires training in its first three phases.

While regulations have historically established minimum standards, companies that want to achieve excellence in safety and health must set their own corporate standards as well. They must determine whether a problem is attributable to training, identify characteristics of effective training, evaluate training programs, implement best training practices, and arrange the training process.

Once a company has defined its training standards, it can select training that will meet its needs, including instruction in safety and health management, orientation, safety and health techniques, and task training.

Finally, companies need to require post-training activities, such as periodic follow-up observations, contacts, and retraining, to ensure that performance change is achieved. These activities promote the transfer of learning from the classroom to the job, which is where companies will see a return on their training investment.

Element 10: Organizational Communications

Executive summary

Companies communicate various kinds of safety and health information. Needs can vary— from internal requirements for progress reports on safety goals and official summaries of safety meetings to informal communication between departments and formal communication from senior management to the entire organization. Management policies are important to communicate. A safety and health policy statement is a key communication tool that shows management's commitment and involvement in the process of improvement.

Organizations may find it desirable to communicate with audiences other than their own employees. Establishing communications with fire and police departments and healthcare providers fosters effective working relationships.

This chapter explores different forms of communication that enhance understanding within the company and between the company and the larger community.

Different types of communications examined here are oral, written, actions, and mechanical devices. Filters that hamper understanding include knowledge, biases, moods, and physical and mental limitations.

The audience is critical to effective communications, but too often its needs are overlooked. Audience requirements affect the kinds, amount, format, and timeliness of information.

Element 11: Management and Control of External Exposures

Executive Summary

Effective management and control of external exposures are an integral part of a company's successful safety and health program, and are key to its overall plan for risk management. Senior management needs to consider issues related to five major kinds of external exposure that can put the company at risk for potential liability.

Contractors, vendors, products produced by the company, public liability, and natural disasters can expose the company to incidents, injuries, and illnesses, as well as to substantial financial loss. Management should establish policies for addressing each exposure.

Companies can take proactive steps to assess and prioritize risks. Techniques for dealing with them include prequalifying contractors and vendors, monitoring changes in standards and regulations (especially in the global marketplace) and the impact of those changes, and adopting a proactive approach to minimizing exposure.

Element 12: Environmental Management

Executive Summary

Safety and health and environment management have many similarities. They all focus on identifying and controlling hazards. Whereas safety and health management addresses hazards to people in the workplace, environmental management broadens the focus to people and other living things in the surrounding community and in other places affected by the organization's operations.

Regulatory compliance issues often drive environmental programs. Federal, state (or regional), and local governments in the United States, the European Union, Canada, and Mexico have environmental regulatory requirements. In addition, there are numerous international environmental agreements in effect. Effective environmental management requires staying aware of, understanding, and complying with these diverse and sometimes conflicting requirements.

As with safety and health management, effective environmental management requires careful assessment of environmental hazards and their solutions followed by prioritization and action. This process involves evaluating all business operations for existing and potential adverse environmental effects and compliance issues, identifying solutions to eliminate or minimize the impact, and, where feasible, implementing them.Element 13: Work Force planning and staffing

Element 13: Workforce Planning and Staffing

Executive Summary

Workforce planning and staffing begin with hiring and job placement. Companies can incorporate safety and health principles into their hiring and placement processes by specifying safety and health requirements, such as indicated in the job description and established by a Job Hazard Analysis, and followed by administering physical examinations and health questionnaires. Many companies' new-employee orientation programs introduce basic safety and health concepts and provide specific information on required subjects such as hazard communication procedures.

Safety and health work rules, or safe work requirements, are usually part of a company's safety and health program. Keys to the success of work rules include employee involvement in their development, effective communication, and consistent enforcement.

Employee Assistance Programs (EAPs) are one way employers address personal problems that can affect an employee's ability to perform on the job. These programs can be an important part of a safety and health effort.

Under the Americans with Disabilities Act (ADA), employers are responsible for making reasonable accommodations for individuals with disabilities. Employers are also responsible for seeing that their workplace is accessible for individuals with disabilities. Because some organizations will have to make changes in job descriptions as well as workplaces, the act may have a large impact on workforce planning and staffing.

Element 14: Assessments, Audits, and Evaluations

Executive Summary

Knowing where and how specific safety policies and programs are succeeding or failing is crucial to continuous improvement. This chapter examines how to implement systems that can provide management with constant and meaningful data on the effectiveness of the safety and health field.

Thorough and objective evaluation of the overall safety and health program requires a variety of evaluative tools. The evaluation process must meet the individual needs of each level of the organization. We examine the broad areas of concern at each level, as well as the types of assessments that address those concerns.

Self-assessment processes utilize trained internal staff to conduct inspections and other ongoing audits. Third-party processes employ consultants to execute comprehensive or complex assessments. Voluntary regulatory assessments involve the cooperative effort of the organization and a regulatory agency to improve safety and health in the workplace. We discuss the advantages and disadvantages of each method.

Any assessment process is incomplete until its findings have been reported to and acted upon by management in a timely and meaningful manner. Management must establish standards for assessment reports and procedures for follow-up that facilitate continuous improvement.

Safety Statistics

We are constantly bombarded with statistics about workplace injuries, fatalities, and the associated costs. At some point we must step away from the statistics that we know about the workplace and look at the injuries and losses away from work. The numbers associated with unintentional deaths and injuries are staggering. If an injury takes place at work or away from work, the employer has to pay in one form or another. As a nation, can we improve the statistics regarding injuries and illnesses away from the workplace if we pay more attention to the job each worker is performing? Can we improve the safety performance at the various worksites throughout the country?

The safety performance and statistics that we enjoy today came about by improving work processes and preventing injuries that occurred to others. Fortunately, we sometimes learn from our mistakes and go forward to keep from repeating the same error. It is important to reflect back on safety progress and note the sad performance of industry around the turn of the century and the improvements along the way. This chapter also identifies the types of injuries and statistics that occur in the workplace as reported by the Bureau of Labor Statistics, the National Safety Council, and several other organizations.

Workers' Compensation Costs

Injuries are costly to business and industry. There is the cost of the premiums paid out for insurance. There is the cost of paying indemnity to those workers that are injured and are losing time from the job. Then there are medical costs, and the costs associated with the worker being absent from a job that requires his skills. There are also numerous indirect costs, such as time spent investigating and reporting the claim, the work stoppage that takes place when an injury occurs, and any property or product damage associated with the incident?

All of the aforementioned factors placed a financial burden on the employer. But we cannot forget the burden placed on the worker and his family. When one is injured, their paycheck shrinks. If the injury was serious, they could be permanently disabled, and they may never be able to return to the job.

The true cost to the nation, to employers, and to individuals of work-related deaths and injuries is much greater than the cost of workers compensation alone, according to the National Safety Council. The following estimates of injuries and fatalities are:

The total cost in 1998 was $122.6 billion. This would include wage and productivity losses of $63.9 billion, medical costs of $19.9 billion, and administrative expenses of $23.5 billion. Included in the totals are employer costs of $11.0 billion, such as the monetary value of time lost by workers other than those with disabling injuries, who are directly or indirectly involved in injuries, and the cost of time to investigate injuries, write up injury reports, etc. Also included in the data are the costs associated with motor vehicle damage, at $2.0 billion, and fire losses, at $2.3 billion.

The cost per worker was $910. This figure indicates the value of goods and services each worker must produce to offset the cost of work injuries. The figure is not the average cost of a work injury. The cost of a worker's death was estimated to be $940,000. The cost of a disabling injury was estimated to be $28,000 and includes wage losses, medical expenses, administrative expenses, and employer costs. Excluded from the total cost are property damage costs except to motor vehicles.

The total time lost in 1999 as a result of work injuries was 80,000,000 days. This figure includes primarily the actual time lost during the year from disabling injuries, except that it does not include time lost on the day of the injury or time required for further medical treatment or check-up following the injured person's return to work.

Fatalities are included in the total time lost as an average of 150 days per case, and permanent impairments are included at actual days lost plus an allowance for lost efficiency resulting from the impairment. Not included is time lost by persons with non-disabling injuries or other persons directly or indirectly involved in the incidents.

The estimated time lost due to injuries in prior years was 45,000,000 days. This is an indicator of the productive time lost in 1999 due to permanently disabling injuries that occurred in prior years. Time lost in future years from 1999 injuries is 60,000,000 days. For any time lost in future years due to on-the-job deaths and permanently disabling injuries, this number is projected to have an impact on time away from work.

A recent survey of workers' compensation insurance carriers indicates that they are planning at least a 20 percent raise in premium rates for the year 2001. For almost the last decade, employers have experienced some relief from high premium charges. This cycle is changing, but it is not in a crisis stage at this time. The National Council on Compensation Insurance, NCCI, which collects workers' compensation and recommends rates and loss costs in 39 states, is looking at rising rates in industry.

The combined rate, which measures the relationship of premium to loss and administrative expenses, has been growing the past four years. For accident year 1999, the combined ratio is 134.6, according to the NCCI. This means that insurers pay $1.35 in expenses for every $1.00 they collect. The NCCI's figures measure lost-time claims, which represent over 90 percent of claims expenses.

A spokesperson from NCCI reported that the average cost per claim has been increasing over the past five years, but this has been offset by a declining number of claims per 100 workers. In 1999/2000, claims frequency remains lower than it was in the 1980's, but the extent of the decrease is tapering off. For accident year 1998, the average cost per lost-time workers' compensation claim was $21,173 compared to $18,183 for accident year 1993, according to NCCI. This represents a 16 percent increase. Most of these higher costs are for medical treatment.

The National Safety Council offered the following information in their Injury Facts 2000 Edition. The data in Table 9-1 are from the National Council on Compensation Insurance Detailed Claim Information file. The data are a stratified random sample of lost time claims in 41 states. Total incurred costs consist of medical and indemnity payments plus case reserves on open claims and are calculated as of the second report (18 months after the initial report of injury). Injuries that result in medical payments only, without lost time, are not included.

Table 9-1 Average Total Incurred Costs Per Claim By Cause Of Injury, 1997-1998

Motor Vehicle	$21,613
Burn	13,361
Fall/Slip	11,670
Miscellaneous Cause	10,500
Cumulative Trauma	10,310
Caught In or Between Object/Equipment	9,222
Strain	9,121
Struck by	8,984
Striking Against	7,613
Cut/Pinch/Scrape	6,624

The most costly lost-time workers' compensation claims by nature of the injury are those resulting from amputation. These injuries averaged $16,468 per workers compensation claim filed in 1997 and 1998. This data are shown in Table 9-2.

Table 9-2 Average Total Incurred Costs Per Claim By Nature Of Injury, 1997,1998

Amputation	$16,468
Fracture/Crush/Dislocation	13,554
Other Trauma	12,692
Carpal Tunnel	11,944
Burn	10,575
Occupational Disease/Cumulative Injury	9,389
Laceration/Puncture/Rupture	9,256
Infection/Inflammation	8,721
Contusion/Concussion	8,328

When viewed by the part of the body, the most costly lost-time workers' compensation claims are for those involving the head or central nervous system. The average cost of these injuries in 1997 and 1998 were $25,650. This information can be seen in Table 9-3.

Table 9-3 Average Total Incurred Costs Per Claim By Part Of Body, 1997-1998

Head/Central Nervous System	$25,650
Multiple Body Parts	17,398
Neck	5,448
Leg	13,688
Knee	11,007
Arm/Shoulder	10,855
Lower Back	10,450
Hip/Thigh/Pelvis	9,970
Upper Back	9,296
Multiple Trunk/Abdomen	8,128
Face, Teeth, Mouth and Eyes	6,606
Ankle	6,528
Chest/Internal Organs	6,503
Hand/Finger/Wrist	6,393
Foot and Toe(s)	5,961

Occupational Health

According to the BLS, Approximately 391,000 occupational illnesses were recognized or diagnosed in 1998. Disorders associated with repeated trauma were the most common illness with 253,000 new cases, followed by over 53,000 cases of skin diseases and disorders. Respiratory conditions due to toxic agents produced 17,500 cases.

The overall incidence rate of occupational illness for all workers was 44.2 per 10,000 full-time workers. Within industry, manufacturing had the highest rate in 1998, 125.5 per 10,000 full-time workers. Workers in manufacturing had the highest rates for disorders associated with repeated trauma, respiratory conditions due to toxic agents, disorders due to physical agents, and poisoning. Agriculture had the second highest incidence rate, 30.9, although agricultural workers had the highest rate of all the industrial divisions for skin diseases and disorders. The mining industry had the highest incidence rate for dust diseases of the lungs.

Back Injuries

Approximately 27 percent of all injuries in the private sector involve the back. The service sector had the highest proportion of injuries involving the back with 31 percent, followed by transportation and public utilities with 28 percent and wholesale and retail trade with 26 percent. The services and manufacturing industries had the highest proportion of injuries to the entire back of the worker, including injury to the spine and spinal column with 27 and 23 percent respectively. The services and manufacturing industries also had the highest proportion of injuries to the lumbar region of the back with 27 and 24 percent respectively, and multiple back regions, 38 and 26 percent respectively. For all industries, injuries to the lumbar region of the back were the most numerous.

As expected in cases of lower back injuries, overexertion in lifting was the leading cause—accounting for almost 75 percent of all back injuries in 1995. This number includes the spine and spinal column. Falls to the same level produced 10 percent of the back injuries while slips or trips without falling produced 7 percent of the totals. These two events were the next highest injury-producing exposures. These three listed exposures were also involved in 94 percent of the injuries to the lumbar region and 90 percent of those to multiple back regions, with overexertion in lifting the leading exposure in injuries to both regions, at 79 and 71 percent respectively.

Sprains and strains accounted for the most injuries to the entire back, at 84 percent and the lumbar region, at 86 percent. In addition, sprains and strains were involved in almost 75 percent of injuries to multiple back regions.

The major sources of injury that caused the greatest number and incidence of back injuries included handling containers, worker motion or body position while at work, walking surfaces, parts and materials, and healthcare patients. Handling containers were associated with over 25 percent of the injuries to the entire back, nearly 25 percent of the injuries to the lumbar region, and almost 18 percent of the injuries to multiple back regions. Worker motions or body position while on the job accounted for between 10 to 18 percent of the injuries to different regions of the back, except for the coccygeal region.

About 25 percent of all back injuries resulted in 21 or more lost workdays. An identical portion of the injuries to the lumbar region resulted in 21 or more lost workdays, while another 25 percent of the injuries to this region resulted in 3-5 lost workdays. Nearly 25 percent of injuries to multiple back regions resulted in 21 or more lost workdays. The lumbar region of the back was the most commonly affected region across the different number of workday losses.

For those readers that would like to maintain control of their workers' compensation costs, the following list was offered by UWC (Strategic Services on Unemployment and Workers' Compensation).

Your "To-Do List" for Workers' Compensation

Employers needn't be caught off-guard by premium and cost increases for workers' comp, but they do need to prepare. Any or all of the following will help you weather a storm:

1. Research and document your company's experience since the last time workers' comp costs began to increase. The last cycle can give you a good indication of what to expect.

2. Find out what strategies were used to contain costs and improve efficiency and re-evaluate them. Establish internal benchmarks, and get comparable benchmarks from other companies.

3. Consider how your company has changed since the last crisis. With less experienced workers, new jobs and possible changes to the manner of work, your company's risk exposure might have changed. Telecommuting, for example, requires a different loss control strategy, because the employer has little control over the safety of the home workplace and how work there is performed.

4. Make sure your safety program is effective and up-to-date. Safety consultants can be hired or obtained through your insurance company.

5. Coordinate management of workers' comp and nonoccupational disability claims to eliminate duplicate claims and assure employees receive timely assistance.

6. Update your claims reporting systems to reduce administrative costs.

7. Monitor claims performance at least monthly to get an early warning of cost increases.

8. Consider charging workers' comp costs to departments so managers understand that they have a direct impact on the bottom line.

9. Train your managers to know how to respond when an employee sustains a work-related injury. Workers are more likely to return to work promptly when they know their bosses care, so managers should stay in touch with their injured employees. This also reduces the likelihood that routine claims will turn into costly "problem" claims.

10. Treat your employees well. Besides boosting productivity, this is the best way to control workers' comp costs. Satisfied workers are less likely to malinger after job injuries or to file phony claims.

11. Get involved when new legislation and regulations threaten to make workers' comp more expensive.

12. Talk to your insurance company or broker and make sure the insurance arrangement you have is best suited to your company's needs".

Bureau of Labor Statistics Data

The Bureau of Labor Statistics (BLS), reported that there were 6,023 fatal injuries in industry in 1999. This number was the same as that of 1998, but 1999 showed an increase in employment. Decreases in job-related deaths from homicide and electrocutions in 1999 were offset by increases from workers being struck by falling objects or being caught in running machinery. Construction work reported the largest number of fatal injuries for any industry and accounted for one-fifth of the total for fatalities.

The manner in which workplace fatalities occurred in 1999 were:

- Highway incidents   1,491
- Falls   717
- Homicides   645
- Struck by object   585
- Air, water and rail transport 385
- Worker struck by vehicle   377
- Non-highway transport   353
- All other events   1,470

On average, 17 workers were fatally injured each day during 1999. Eighty-three percent of fatally injured workers died the same day they were injured; 97 percent died within 30 days. There were 235 multiple fatality incidents (incidents that resulted in two or more worker deaths), resulting in 617 job-related deaths. Although this was a slight increase over the 227 multiple-fatality events reported for 1998, there was a more substantial increase in the number of deaths resulting from these types of incidents in 1999 than in the previous year, when 555 worker deaths occurred.

The data can be confusing to the reader. Different sources sometimes offer modified or different statistics to make their point. A report from the National Safety Council cites 5,100 workplace fatalities versus the 6,023 cited by the BLS. In another article NIOSH is quoted as stating that 5,840 fatalities occurred in 1999. Occupational disease also generates varied numbers on fatalities. NIOSH reports that 137 workers die each day from work-related diseases; that is more than 50,000 fatalities each year. In many cases, occupational diseases do take their toll on workers, but this tragedy goes mostly unreported in the press.

Information on costs of injuries and illnesses can also be confusing. One report cited a cost to the country as $122.6 billion. Another report offers a direct cost of $65 billion plus indirect costs of $106 billion. Workplace-related costs were estimated to be $171 billion with $145 billion for injuries and $26 billion for illnesses.

All four of the occupational injury and illness incidence rates published by the BLS for 1998 decreased from 1997. The incidence rate for total nonfatal cases was 6.7 per 100 full time workers in 1998 and was down 6% from the 1997 rate of 7.1. This rate was over 11 in 1972. This represents another measurement of the effect OSHA has had on workplace injuries. The 1998 incidence rate for total lost workday cases was 3.1, down 6% from 3.3 in 1997. The incidence rate for lost workday cases with days away from work was 2.0 in 1998. This rate was down 5% from 2.1 in 1997. The incidence rate in 1998 for nonfatal cases without lost workdays was 3.5, a decrease of 8% from the 1997 rate of 3.8.

OSHA's Injury Statistics

Many feel that OSHA has had a real impact on injury and illness reduction in the workplace. History shows that in the early 1970's, when OSHA started completing workplace inspections, workers suffered over 14,000 fatalities each year. Prior to this period of time, the highest recorded fatality count in the workplace took place in 1937 and some 19,000 were killed. The death rate per 100,000 workers has declined from its highest number in 1937 of 43.0 to 3.8 in 1999. In 1937 there were 44,100,000 workers as compared to 134,688,000 in 1999. This data does a favorable job of highlighting the value and programs of OSHA.

What are the types of incidents that result in worker fatalities? Note the sources of fatal occupational injuries in Table 9.4 and non-fatal occupational injuries involving days away from work in Table 9.5 as reported by the BLS.

Table 9-4, Fatal Occupational Injuries, 1998

(of 6026 Fatalities)

Highway Accidents	23.7 %
Assaults	11.8
Falls to a Lower Level	10.3
Struck-by Object	8.6
Pedestrian	6.9
Non-highway Accident	6.4
Electric Current	5.5
Caught-in Equipment	4.4
Aircraft Accident	3.7
Self Inflicted	3.7
All Other Events	15.0

Highway traffic incidents were the leading cause of all work-related fatal injuries in 1998 and the leading cause of these fatalities were in the Transportation and Public Utilities, Services and Government industries. Assaults and violent acts by persons (homicide) was the second leading cause overall and the leading cause in the Wholesale and Retail Trade industry. Falls to a lower level was the third leading cause for all industries and the leading cause in the construction industry.

Table 9.5, Nonfatal Occupational Injuries and Illnesses, Including Days Away From Work, 1998

Contact With Objects and Equipment	27.6%
Overexertion	27.6
Fall Same Level	10.7
Fall Lower Level	5.5
Exposure	4.7
Transportation	4.0
Repetitive Motion	3.8
Slips, Trips	3.2
Assault	1.0
All Others	11.7

More than half of the 1,730,500 non-fatal occupational injuries in private industry involving days away from work result from contact with objects or equipment or from overexertions, according to the BLS 1998 Survey of Occupational Injuries and Illnesses.

Some 102 million workers will be covered by OSHA's proposed ergonomic standard. Data show that more than 600,000 workers miss at least one day each year due to injuries associated with repetitive stretching, bending, lifting or typing. OSHA states that over 1.8 million workers suffer from ergonomic injuries. [CITE]

OSHA contends that industry spends $15 billion to $20 billion each year in compensation related to musculoskeletal disorders (MSDs). This outlay is one-third of all workers' compensation costs. OSHA has predicted that the proposed ergonomics rules would cost businesses some $4.5 billion to implement, but would reap $9 billion a year in savings from medical expenses and workers' compensation costs. The actual cost to industry has been a point of contention with several business groups. With OSHA's position on costs, calculated at an average of $250 per worker station in offices and factories, the National Association of Manufacturers argues that the real costs will be $781 per employee, or $6.7 billion in the first year. But, says OSHA, the ergonomics standard could slash repetitive strain incidents by 75 percent.

Results of OSHA's Inspections

OSHA continues to conduct workplace inspections and cite employers that violate OSHA standards. In fiscal 2000, OSHA's programmed inspections climbed 18 percent from the number of inspections conducted in fiscal 1999. The number of planned inspections rose to 18,324 from 15,530 conducted in the previous year. Total inspections rose five percent to 36,202 in FY 2000 from 34,488 in FY1999.

Employers were cited for violating 80,494 standards in FY 2000. This is a six percent rise from 75,989 in FY 1999. Penalties assessed rose some 18 percent to $86.9 million from $73.6 million in FY 1999.

What are the items that employers are cited for during the OSHA inspections? This is a frequently asked question. The following top 10 violations would have been in the willful, serious, and repeat categories. Employers are urged to establish sound safety programs that prevent workplace injuries and illnesses. This process takes effort and review as well as management commitment. The listing can be noted in Figure 9-6.

Table 9-6 OSHA's Top 10 Violations

1. Scaffolding—Construction 29 CFR 1926.451
2. Fall Protection—Construction 29 CFR 1926.501
3. Hazard communication—General Industry 29 CFR 1910.1200
4. Machine Guarding—General Industry 29 CFR 1910.212
5. Lockout/Tagout—General Industry 29 CFR 1910.147
6. Respiratory Protection—General Industry 29 CFR 1910.134
7. Electrical,wiring, components, methods and equipment—General Industry 29 CFR 1910.305
8. Mechanical Power, Transmission and Apparatus—General Industry 29 CFR 1910.219
9. Mechanical Power Presses—General Industry 29 CFR 1910.217
10. Electrical Systems Design—General Industry 29 CFR 1910.303

Comments on the above 10 violations should provide readers with some background information on each of the items.

- The scaffolding standard (1926.451) requires employers to correctly install scaffolds that include hand railing, planking for the walking and working surface, security for the base posts, protection from being struck by vehicles, and security to prevent any shifting or tipping. A JHA could easily be developed to identify the basic steps, hazards, and the exact method of safely erecting a scaffold. For the most part, once the JHA was completed, it could be used at multiple job sites with just a few revisions. The process of scaffold erection at any job site should contain the same safety requirements and basic steps.

- The fall protection standard (1926.501) relates to fall hazards at construction sites that are not related to scaffolding. Projects such as roofing, erecting structural steel, painting, working from ladders or cranes, and any other situation that could cause a fall are a part of those violations that result in a penalty. Falls account for 10.3 percent of all fatal injuries in the workplace, and are the leading cause of death in construction. It should be noted that some of these same hazards exist in general industry and must be eliminated to protect all workers. A JHA can be developed for each of the workplace hazards identified.

- Violations of the Hazard Communication Standard (1910.1200) continue to be the most frequently cited violations. The violations are in the serious, willful, or repeat category. It should be pointed out that this standard dates back to 1986, and it continues to pose a problem for many industries. Briefly, the standard requires employee training, material safety data sheets, and a comprehensive list of chemicals; container labeling; and employee training. One recent OSHA citation was for $155,000 for not training employees about the use of hazardous chemicals.

- Machine guarding (1910.212) citations are the result of an injury to a worker. Several chapters of this book go into many details regarding the problems as well as the solutions to machine guarding problems. Employers should be reminded that machine guarding injuries do not occur in a facility, but when they do occur, they can be very serious.

- The OSHA lockout/tagout standard (1910.147) has been with the OSHA standards since day one. In 1991, OSHA developed a formal standard for all of industry to prevent some 140 deaths from occurring each year. This standard requires: employee training, a written program, hardware to use for locking and tagging, and a comprehensive listing of all the operations and devices that need to be locked out, how they will be locked out, and who conducted the hazardous energy survey. One organization was fined $105,000 for violating the standard in fiscal 2000.

- Respiratory protection (1910.134) continues to be at the top of OSHA's most violated standards. This standard requires a concentrated focus to ensure compliance with the various details that OSHA has developed. A proposed penalty of $118,000 was given to one firm after a worker died at a chemical plant.

- Non-compliance with the National Electric Code continues to be a major source of citations as a result of violating 1910.305, OSHA's wiring methods, components, and equipment under the electrical standard. One employer was cited for 35 serious violations that resulted in a penalty of $128,000.

• Mechanical power, transmission apparatus (1910. 219). Included in this standard are devices that transmit power such as drive shafts, pulleys and pulley belts, chains, sprockets, belts, and gears. These dangerous devices require guarding. One employer was cited for $158,000 for failing to guard pulleys and assorted other devices.

• Mechanical power presses (1910.217) is a part of OSHA's national emphasis program. Without machine guarding on presses, a worker is exposed to some very serious hazards. The number of citations written by OSHA for the lack of guarding on insufficient guarding propelled this standard from 22nd in 1997 to fifth in 1998. One organization received a penalty of $67,200 for power press violations.

• Electrical systems design (1010.303) covers the guarding of live parts on machinery. The main hazard covered under this standard is electrocution.

Agenda for the Nation

Jerry Scannell, former President of the National Safety Council, NSC, and former head of OSHA during the Bush administration, has provided the nation with an agenda to reduce unintentional deaths. This document was developed through the cooperation of business, labor, safety and health professionals, and the government. The National Safety Council stated that each year, unintentional injuries kill more Americans between one and 44 years old than any other cause, including cancer, heart disease and stroke. In 1999, unintentional injuries resulted in:

• 95,500 deaths—the highest death toll since 1988.

• The disabling of 20.4 million people for a day or more.

In 1998 unintentional injuries resulted in:

• 28 million hospital emergency room visits, and more than 56 million visits to a physicians offices.

• 54 million people having 598 million days of restricted activity for a day or more.

Do people get used to tragedy? Do they become so conditioned to seeing reports of the individuals that are injured or killed on the highway, at work, and at home that they pay no heed to these reports? The NSC also finds it disturbing that the enormous cost of these tragedies, $500 billion in 1999, is unknown to many. For the most part, these losses are preventable.

In the home and community, approximately 14,500,000 people suffered disabling injuries at home and public places in 1999. There were 52,000 injury deaths, a 21 percent increase from the 43,000 deaths in 1992. While unintentional deaths were up 21 percent, the nation's population rose 7 percent from 1992 to 1999. That statistic gives an increase in the death rate in the home and community of 19.1 percent per 100,000 population.

On the highway, the peak year for unintentional injury deaths was 1972. For that year, the rate of highway deaths was 26.9 per 100,000 population. In 1999, the rate was reduced to 15.0 deaths. In 1982, 56,278 people died on the highways compared to 40,800 in 1999.

The drop in the death rate is attributable to safer cars, better highway design, seat belts and strong enforcement of drunk driving laws. Even though the trend indicates fewer highway deaths per 100,000 population, the losses that do take place must be reduced to their lowest level possible.

In the workplace, there has been great progress toward injury reduction. However, there is much more to do to reduce on-the-job highway fatalities, falls in the construction industry, repetitive stress injuries, occupational diseases, and the overall injury rate.

To aid in the solution of injury reduction in the workplace, the NSC recommends that changes be made to the safety process. Implementing a systematic approach to workplace safety will require a cultural change in many organizations. All organizations need to nurture a safety culture. Company policy and workstation practice must dictate that safety never take a back seat to other interests. No one should be asked, and no one should tolerate, a potentially disabling or life-threatening risk in the name of cost-cutting, productivity, or any other priority.

Safety and health considerations must be an integral part of the operating policies of every organization. The consequences are too expensive when safety and health are relegated to a position of just one of many changing priorities.

The NSC asks senior management of private and public enterprises to institute and communicate to all employees a Corporate Code of Safety and Health Ethics. The Code will:

- Add safety and health to the core values of the organization, on the same level and with the same support as other values defined by the organization, such as customer service, financial performance, and productivity.

- Recognize that the Chief Executive Officer (CEO) or public equivalent is the leader of the organization, and all of his or her actions and decisions add to or detract from the credibility of the safety and health commitment.

- Ensure that the CEO and senior management personally set the standard for safety and health performance.

- Establish a comprehensive evaluation, a safety and health audit of the organization, to identify existing and evolving hazards and to develop an effective system of accountability to ensure that they are controlled or abated.

- Recommend that the corporation's board meeting and annual report to shareholders include a discussion of the company's safety and health performance.

A Concern with Literacy

Some industry leaders are concerned with the growing rate of illiteracy among workers. The American Management Association, AMA, reported that 41.7 percent of manufacturing job applicants have basic skills deficiencies. More than 38 percent of all applicants lack the necessary reading, writing, and math skills to do the jobs they seek. In 1999, the share of skills-deficient job applicants was up from 35.5 percent in 1998, and 22.8 percent in 1997, reports the AMA. Are factories running out of qualified individuals to perform the necessary work? Could there be a correlation between individuals who are injured on and off the job and the amount of skills and education they possess?

The AMA stated, "for all the concern about the shortage of high-tech workers, industry is facing a lack of job applicants who can read and write. Companies are going to have to find new ways to staff themselves with qualified workers. They are not going to be able to rely merely on selective hiring to achieve their goals. They will have to invest more in training new hires and current employees. In many cases, businesses are going to be compelled to develop rather than hire workforces."

If literacy is an issue and a growing one at that, methods will have to be devised to properly train these workers that may lack the skills to understand and apply safety tasks with the job being performed. Of course, the use of job safety analysis can do much to develop a safe workplace and a safe worker. A key to working safely is the need to know and understand safety rules and guidelines; especially those that are a part of critical job applications such as operating a machine and confined space hazards.

Safety and Health Resources

An entire book can be written on what should be contained in a safety and health program. Many corporations maintain excellent safety and health programs, but this success did not come easy. It takes a considerable amount of planning, dedication, sacrifice, and energy to achieve excellence in protecting workers and maintaining a successful safety and health program.

For those readers that want to improve their programs and reduce workers' compensation costs, there is a significant amount of information available in a variety of forms and from many different sources. The following sources provide assistance and information to businesses seeking to improve their safety and health programs:

OSHA regional and local offices (see Appendices B and C):

The National Safety Council (NSC)
1121 Spring Lake Drive
Itasca, IL 60143-3201
(630) 285-1121
(630) 285-1315
http://www.nsc.org

The American Society of Safety Engineers (ASSE)
1800 East Oakton Street
Des Plaines, IL 60018
(847) 699-2929
(847) 768-3434 (fax)
http://www.asse.org

The American Industrial Hygiene Association (AIHA)

2700 Prosperity Avenue, Suite 250

Fairfax, VA 22031-4307

(703) 849-8888

(703) 207-3561 (fax)

http://www.aiha.org

Local safety councils, many of which are chapters of larger organizations such as the National Safety Council.

Professional organizations that assist in developing safety standards for the country, including:

National Fire Protection Association (NFPA)

One Batterymarch Park

P.O. Box 9101

Quincy, MA 02269-9101

(617) 770-3000

(617) 770-0200 (fax)

http://www.nfpa.org

American National Standards Institute (ANSI)

1819 L Street, NW

Washington, DC 20036

(202) 293-8020

(202) 293-9287 (fax)

http://www.ansi.org

Independent consultants familiar with your specific industry.

Trade or industry associations (a particularly good source because the information is usually relevant to individual operations, which saves the time of plowing through extraneous standards and programs).

Appendix B

OSHA Regional Offices

In case of emergency, call 1-800-321-OSHA.

Region 1
CT, MA, ME, NH, RI, VT
JFK Federal Building, Room E340
Boston, MA 02203
Telephone: (617) 565-7164
Fax: (617) 565-9827

Region 2
NJ, NY, Puerto Rico, Virgin Islands
201 Varick Street, Room 670
New York, NY 10014
Telephone: (212) 337-2378
Fax: (212) 337-2371

Region 3
DC, DE, MD, PA, VA, WV
Gateway Building, Suite 2100
3535 Market Street
Philadelphia, PA 19104
Telephone: (215) 596-1201
Fax: (215) 596-4872

Region 4
AL, FL, GA, KY, MS, NC, SC, TN
61 Forsyth Street, SW
Atlanta, GA 30303
Telephone: (404) 562-2300
Fax: (404) 562-2295

Region 5
IL, IN, MI, MN, OH, WI
230 South Dearborn Street, Room 3244
Chicago, IL 60604
Telephone: (312) 353-2220
Fax: (312) 353-7774

Region 6
AR, LA, NM, OK, TX
525 Griffin Street, Room 602
Dallas, Texas 75202
Telephone: (214) 767-4731
Fax: (214) 767-4137

Region 7
IA, KS, MO, NE
1100 Main Street, Suite 800
Kansas City, MO 64105
Telephone: (816) 426-5861
Fax: (816) 426-2750

Region 8
CO, MT, ND, SD, UT, WY
1999 Broadway, Suite 1690
Denver, CO 80202-5716
Telephone: (303) 844-1600
Fax: (303) 844-1616

Region 9
AZ, CA, Guam, HI, NV
71 Stevenson Street, Room 420
San Francisco, CA 94105
Telephone: (415) 975-4310
Fax: (415) 975-4319

Region 10
AK, ID, OR, WA
1111 Third Avenue, Suite 715
Seattle, WA 98101-3212
Telephone: (206) 553-5930
Fax: (206) 553-6499

Appendix C

OSHA Consultation Services

Personnel within these offices will provide assistance and information relevant to that particular state as well as federal requirements. The states are listed in alphabetical order.

Alabama
Safe State Program
University of Alabama
432 Martha Parham West
P.O. Box 870388
Tuscaloosa, AL 35487
(205) 348-3033
(205) 348-3049 (fax)
bweems@ccs.ua.edu
http://bama.ua.edu/~deip/safest.html

Alaska
Consultation Section, ADOL/AKOSH
3301 Eagle Street
P.O. Box 107022
Anchorage, AK 99510
(907) 269-4957
(907) 269-4950 (fax)
timothy_bundy@labor.state.ak.us
http://www.labor.state.ak.us/lss/oshhome.htm

Arizona
Consultation and Training
Industrial Commission of Arizona
Division of Occupational Safety and Health
800 West Washington
Phoenix, AZ 85007-9070
(602) 542-5795
(602) 542-1614 (fax)
henry@n245.osha.gov

Arkansas
OSHA Consultation
Arkansas Department of Labor
10421 West Markham
Little Rock, AR 72205
(501) 682-4522
(501) 682-4532 (fax)
clark@n237.osha.gov
http://www.state.ar.us/labor/serv01.htm

California
CAL/OSHA Consultation Service
Department of Industrial Relations
455 Golden Gate Avenue, 10ᵗʰ Floor
San Francisco, CA 94102
(415) 703-5270
(415) 703-4596 (fax)
InfoCons@hq.dir.ca.gov
http://www.dir.ca.gov/DOSH/consultation.html

Colorado
Colorado State University
Occupational Safety and Health Section
115 Environmental Health Building
Fort Collins, CO 80523
(970) 491-6151
(970) 491-7778 (fax)
rbuchan@lamar.colostate.edu
http://www.bernardino.colostate.edu/enhealth/
7c1.html

Connecticut
Connecticut Department of Labor
Division of Occupational Safety and Health
38 Wolcott Hill Road
Wethersfield, CT 06109
(860) 566-4550
(860) 566-6916 (fax)
donald.heckler@ct-ce-wethrsfld.osha.gov
http://www.ctdol.state.ct.us/osha/osha.htm

Delaware
Delaware Department of Labor
Division of Industrial Affairs
Occupational Safety and Health
4425 Market Street
Wilmington, DE 19802
(302) 761-8219
(302) 761-6601 (fax)
ttrznadel@state.de.us
http://www.state.de.us/labor/aboutdol/
industrialaffairs.html

District of Columbia
(Program available only for employers within
DC)
DC Department of Employment Services
Office of Occupational Safety and Health
950 Upshur Street, NW
Washington, DC 20011
(202) 576-6339
(202) 576-7579 (fax)
jcates3@aol.com

Florida
Florida Department of Labor and Employment
Security
7(c)(1) Onsite Consultation Program
Division of Safety
2002 St. Augustine Road, Building E, Suite 45
Tallahassee, FL 32399-0663
(850) 922-8955
(850) 922-4538 (fax)
colette_drouillard@safety.fdles.state.fl.us
http://www.safety-fl.org/consult.htm

Georgia
Georgia Institute of Technology
7(c)(1) Onsite Consultation Program
151 6ᵗʰ Street, NW
O'Keefe Building, Room 22
Atlanta, GA 30332-0837
(404) 894-2643
(404) 894-8275 (fax)
daniel.ortiz@gtri.gatech.edu
http://www.gtri.gatech.edu/safety.htm

Guam
OSHA Onsite Consultation
Department of Labor, Government of Guam
P.O. Box 9970
Tamuning, GU 96931
011 (671) 475-0136
011 (671) 477-2988 (fax)
http://mail.admin.gov.gu/webdol/
oshaco_mpl.htm

Hawaii
Consultation and Training Branch
Department of Labor and Industrial Relations
830 Punchbowl Street
Honolulu, HI 96813
(808) 586-9100
(808) 586-9099 (fax)
ellen.kondo@hi-e-honolulu.osha.gov
http://www.aloha.net/~edpso/annual.html

Idaho
Boise State University
Department of Health
Safety and Health Consultation Program
Boise, ID 83725
(208) 426-3283
(208) 426-4411 (fax)
Istokes@boisestate.edu
http://www.idbsu.edu/health/healthstudies/

Illinois
Illinois Onsite Consultation
Industrial Service Division
Department of Commerce and Community
Affairs
State of Illinois Center, Suite 3-400
100 West Randolph Street
Chicago, IL 60601
(312) 814-2337
(312) 814-7238 (fax)
sfryzel@commerce.state.il.us
http://www.commerce.state.il.us/Services/
SmallBusiness/OSHA/OSHAhome.htm

Indiana
Bureau of Safety, Education and Training
Division of Labor, Room W195
402 West Washington
Indianapolis, IN 46204-2287
(317) 232-2688
(317) 232-3790 (fax)
jon.mack@in-ce-indianpls.osha.gov
http://www.state.in.us/labor/busets/buset.html

Iowa
7(c)(1) Consultation Program
Iowa Bureau of Labor
2016 DMACC Boulevard
Building 17, Room 10
Ankeny, IA 50021
(515) 965-7162
(515) 965-7166 (fax)
sslater@n192.osha.gov
http://www.state.ia.us/iwd/labor/index.html

Kansas
7(c)(1) Consultation Program
Kansas Department of Human Resources
512 South West 6th Street
Topeka, KS 66603-3150
(785) 296-7476
(785) 296-1775 (fax)
rudy.leutzinger@ks-c-topeka.gov

Kentucky
Kentucky Labor Cabinet
Division of Education and Training
1047 U.S. Highway 127, South
Frankfort, KY 40601
(502) 564-6895
(502) 564-6103 (fax)
arussell@mail.lab.state.ky.us

Louisiana
7(c)(1) Consultation Program
Louisiana Department of Labor
1001 North 23rd Street, Room 230
P.O. Box 94094
Baton Rouge, LA 70804-9094
(225) 342-9601
(225) 342-5158 (fax)
sandra.purvis@la-c-batonrouge.osha.gov

Maine
Division of Industrial Safety
Maine Bureau of Labor Standards
Workplace Safety and Health Division
State House Station #45
Augusta, MA 04333-0045
(207) 624-6460
(207) 624-6449 (fax)
david.e.wacker@state.me.us
http://janus.state.me.us/labor/consult.htm

Maryland
MOSH Consultation Services
312 Marshall Avenue, Room 600
Laurel, MD 20707
(410) 880-4970
(410) 880-6369 (fax)
virginia.anklin@md-r7-laurel.osha.gov
http://www.dllr.state.md.us/labor/mosh.html

Massachusetts
Division of Occupational Safety and Health
Department of Workforce Development
1001 Watertown Street
West Newton, MA 02165
(617) 727-3982
(617) 727-4581 (fax)
Joe.LaMalva@state.ma.us
http://www.state.ma.us/dos/Consult/Consult.htm

Michigan (Health)
Occupational Health Division
7150 Harris Drive
P.O. Box 30649
Lansing, MI 48909
(517) 322-6823
(517) 322-1775 (fax)
john.peck@cis.state.mi.us
http://www.cis.state.mi.us/bsr/divisions/occ/
occcons.htm

Michigan (Safety)
Department of Consumer and Industry Services
7150 Harris Drive
Lansing, MI 28909
(517) 322-1809
(517) 322-1374 (fax)
ayalew.kanno@cis.state.mi.us
http://www.cis.state.mi.us/bsr/divisions/set/
set_con.htm

Minnesota
Department of Labor and Industry
Consultation Division
443 Lafayette Road
Saint Paul, MN 55155
(612) 297-2393
(612) 297-1953 (fax)
james.collins@state.mn.us
http://www.doli.state.mn.us/mnosha.html

Mississippi
Mississippi State University
Center for Safety and Health
2906 North State Street, Suite 200
Jackson, MS 39216
(601) 987-3981
(601) 987-3890 (fax)
Kelly.tucker@ms-c-jackson.osha.gov
http://www.msstate.edu/dept/csh/

Missouri
Onsite Safety and Health Consultation Service
3315 West Truman Boulevard
P.O. Box 449
Jefferson City, MO 65102
(800) 475-2130 (select option #4)
(573) 751-3721 (fax)
rsimmons@dolir.state.mo.us
http://www.dolir.state.mo.us/ls/onsite/index.html

Montana
Department of Labor and Industry
Bureau of Safety
P.O. Box 1728
Helena, MT 59624-1728
(406) 444-6418
(406) 444-4140 (fax)
jandersen@state.mt.us
http://dli.state.mt.us/publica.htm

Nebraska
Division of Safety and Labor Standards
Nebraska Department of Labor
State Office Building, Lower Level
301 Centennial Mall, South
Lincoln, NE 68509-5024
(402) 471-4717
(402) 471-5039 (fax)
ediedrichs@dol.state.ne.us
http://www.dol.state.ne.us/safety/7c1.htm

Nevada
Safety Consultation and Training Section
Division of Industrial Relations
Department of Business and Industry
1301 Green Valley Parkway
Henderson, NV 89014
(702) 486-9140
(702) 990-0362 (fax)
dalton.hooks@nv-ce-lasvegas.osha.gov
http://www.state.nv.us/b&i/ir/index.htm

New Hampshire
New Hampshire Department of Health and
Human Services
6 Hazen Drive
Concord, NH 03301-6527
(603) 271-2024
(603) 271-2667 (fax)
stephen.beyer@nh-concord.osha.gov
http://www.state.nh.us/dhhs/ohm/dphs.htm

New Jersey
New Jersey Department of Labor
Division of Public Safety and Occupational
Safety and Health
225 East State Street, 8th Floor West
P.O. Box 953
Trenton, NJ 08625-0953
(609) 292-3923
(609) 292-4409 (fax)
carol.farley@nj-c-trenton.osha.gov
http://www.state.nj.us/labor/consult.html

New Mexico
New Mexico Environment Department
Occupational Health and Safety Bureau
525 Camino de Los Marquez, Suite 3
P.O. Box 26110
Santa Fe, NM 87502
(505) 827-4230
(505) 827-4422 (fax)
Debra_McElroy@nmenv.state.nm.us
http://www.nmenv.state.nm.us/env_prot.html

New York
Division of Safety and Health
State Office Campus
Building 12, Room 130
Albany, NY 12240
(518) 457-2238
(518) 457-3454 (fax)
james.rush@ny-ce-albany.osha.gov
http://www.labor.state.ny.us/html/employer/
p469.html

North Carolina
Bureau of Consultative Services
North Carolina Department of Labor—OSHA
Division
4 West Edenton Street
Raleigh, NC 27601-1092
(919) 807-2905
(919) 807-2902 (fax)
wjoyner@mail.dol.state.nc.us
http://www.dol.state.nc.us/osha/consult/
consult.htm

North Dakota
North Dakota State Department of Health
Division of Environmental Engineering
1200 Missouri Avenue, Room 304
Bismarck, ND 58504
(701) 328-5188
(701) 328-5200 (fax)
ccmail.dmount@ranch.state.nd.us
http://www.ehs.health.state.nd.us/ndhd/environ/
ee/oshc/index.htm

Ohio
Bureau of Employment Services
Division of Onsite Consultation
145 South Front Street
Columbus, OH 43215
(614) 644-2246
(614) 644-3133 (fax)
owen@n222.osha.gov
http://www.state.oh.us/obes/osha.htm

Oklahoma
Oklahoma Department of Labor
OSHA Division
4001 North Lincoln Boulevard
Oklahoma City, OK 73105-5212
(405) 528-1500
(405) 528-5751 (fax)
leslie@n238.osha.gov
http://www.state.ok.us/~okdo/osha/index.htm

Oregon
Oregon OSHA
Department of Consumer and Business Services
350 Winter Street, NE, Room 430
Salem, OR 97310
(503) 378-3272
(503) 378-5729 (fax)
steve.g.beech@state.or.us or
consult.web@state.or.us
http://www.cbs.state.or.us/external/osha/consult/
consult2.htm

Pennsylvania
Indiana University of Pennsylvania
OSHA Consultation Service
Walsh Hall, Room 210
302 East Walk
Indiana, PA 15705-1087
(724) 357-2396
(724) 357-2385 (fax)
john.engler@pa-c-indiana.osha.gov
http://www.iup.edu/sa/osha/index.html

Puerto Rico
Occupational Safety and Health Office
Department of Labor and Human Resources
505 Munoz Rivera Avenue, 21st Floor
Hato Rey, PR 00918
(787) 754-2171
(787) 767-6051 (fax)
alopez@n114.osha.gov

Rhode Island
Occupational Safety and Health Consultation
Program
Division of Occupational Health and Radiation
Control
Rhode Island Department of Health
3 Capital Hill
Providence, RI 02908
(401) 222-2438
(401) 222-2456 (fax)
jimg@doh.state.ri.us
http://www.state.ri.us/dohrad.htm

South Carolina
South Carolina Department of Labor, Licensing
& Regulations
3600 Forest Drive
P.O. Box 11329
Columbia, SC 29204
(803) 734-9614
(803) 734-9741 (fax)
bob.peck@sc-c-columbia.osha.gov
http://www.llr.state.sc.us/oshavol.htm

South Dakota
Engineering Extension
Onsite Technical Division
South Dakota State University
West Hall, Box 510
907 Harvey Dunn Street
Brookings, SD 57007
(605) 688-4101
(605) 688-6290 (fax)
letterms@cc.sdstate.edu

Tennessee
OSHA Consultation Services Division
Tennessee Department of Labor
710 James Robertson Parkway, 3rd Floor
Nashville, TN 37243-0659
(615) 741-7036
(615) 532-2997 (fax)
mmaenza@mail.state.tn.us
http://www.state.tn.us/labor/toshcons.html

Texas
Workers' Health and Safety Division
Texas Workers' Compensation Commission
Southfield Building
4000 South I H 35
Austin, TX 78704
(512) 804-4640
(512) 804-4641 (fax)
(800) 687-7080 (OSHCON Request Line)
jharper@twcc.state.tx.us
http://twcc.state.tx.us/services/oshcon.html

Utah
State of Utah Labor Commission
Workplace Safety and Health Consultation
Services
160 East 300 South
Salt Lake City, UT 84114-6650
(801) 530-6901
(801) 530-6992 (fax)
jcmain.nanderso@state.ut.us http://
www.labor.state.ut.us/
Utah Occupational Safety   Hea/
Consultation Services/
consultation services.html

Vermont
Division of Occupational Safety and Health
Vermont Department of Labor and Industry
National Life Building, Drawer 20
Montpelier, VT 05602-3401
(802) 828-2765
(802) 828-2195 (fax)
robert.mcleod@labind.lab.state.vt.us

Virginia
Virginia Department of Labor and Industry
Occupational Safety and Health
Training and Consultation
13 South 13th Street
Richmond, VA 23219
(804) 786-6359
(804) 786-8418 (fax)
njakubecdoli@sprintmail.com
http://www.dli.state.va.us/programs/
consultation.htm

Virgin Islands
Division of Occupational Safety and Health
Virgin Islands Department of Labor
3021 Golden Rock
Christiansted
St. Croix, VI 00840
(340) 772-1315
(340) 772-4323 (fax)
http://www.gov.vi/vild/

Washington
Washington Department of Labor and Industries
Division of Industrial Safety and Health
P.O. Box 44643
Olympia, WA 98504
(360) 902-5638
(360) 902-5459 (fax)
jame235@lni.wa.gov
http://www.wa.gov/lni/wisha/wisha.htm

West Virginia
West Virginia Department of Labor
Capitol Complex Building #3
1800 East Washington Street, Room 319
Charleston, WV 25305
(304) 558-7890
(304) 558-9711 (fax)
jburgess@labor.state.wv.us
http://www.state.wv.us/labor/sections.htm

Wisconsin (Health)
Wisconsin Department of Health and Human
Services
Division of Public Health
Section of Occupational health, Room 112
1414 East Washington Avenue
Madison, WI 53703
(608) 266-8579
(608) 266-9383 (fax)
moente@dhfs.state.wi.us
http://www.dhfs.state.wi.us/dph_boh/
OSHA_Cons/index.htm

Wisconsin (Safety)
Wisconsin Department of Commerce
Bureau of Marketing, Advocacy and Technology
Development
Bureau of Manufacturing and Assessment
N14 W23833 Stone Ridge Drive, Suite B100
Waukesha, WI 53188-1125
(262) 523-3040 or (800) 947-0553
(262) 523-3046 (fax)
jim.lutz@wi-c-waukesha.osha.gov or
wiscon@commerce.state.wi.us
http://www.commerce.state.wi.us/MT/MT-FAX-
0928.html

Wyoming
Wyoming Department of Employment
Workers' Safety and Compensation Division
Herschler Building, 2 East
122 West 25th Street
Cheyenne, WY 82002
(307) 777-7786
(307) 777-3646 (fax)
sfoste1@missc.state.wy.us
http://wydoe.state.wy.us/wscd/osha/evtap.htm

References

Anderson, Joan. "JSA: Your Route to Safety," *Today's Supervisor*, Vol. 57, No.2, February 1993, pp. 10-11.

Armco Steel Corporation. *Accident Prevention Fundamentals and Industrial Hygiene*, Middletown, Ohio, 1976.

Brown, David B. *Systems Analysis and Design for Safety*, Prentice Hall, Englewood Cliffs, NJ, 1976, pp. 42-51.

Bureau of National Affairs. "Ergonomics, OSHA Outlines Rationale for Changes to New Program Standard in Preamble," *BNA Reporter*, November 23, 2000, pp. 1088-1091.

Bureau of National Affairs, Oct. 10, 2000. "Draft of Final Rule for Ergonomics Standard," *BNA Reporter*, November 9, 2000, pp. 1018-1030.

Chadd, Chares, Bowman, Jerome. "OSHA's Discussion Draft of a Safety and Health Standard," Illinois Safety Council Breakfast, April 1, 1997.

Chicago Sun-Times. Worknotes –" Most Injuries on Job Caused by Human Error, Safety Study Finds," Monday, September 6, 1993.

Clark, Kathryn. "Some Organizations are Seeing a Significant ROI on Safety and Health Programs," *Human Resources Executive*, August 2000, p.30.

Colvin, Raymond. *The Guidebook to Successful Safety Programming*, Lewis Publishers, Inc. Chelsea, MI, 1992.

Cullen, Lisa. "Speaking Out-Safety by the (wrong) Numbers?", *Occupational Hazards*, October 2000, p. 145.

De Reamer, Russell. *Modern Safety and Health Technology*, John Wiley and Sons, 1980, pp. 161-164.

Dunn, Richard. Editorial-"The Wheels Are Coming Loose," *Plant Engineering*, July 2000, p.12.

Eninger, M.V. "Managing a Job safety Analysis Program," Bethlehem Steel Corporation, Bethlehem, PA, 1972, pp.37-42.

Eninger, M. V.. "Accident Investigation, Reporting, and Follow-Up," Bethlehem Steel Corporation, Bethlehem, PA, 1972, PP15-28.

Eninger, M.V. "Introduction to Accident Prevention," Bethlehem Steel Corporation, Bethlehem, PA, 1972, pp.1-8.

Eninger, M. V. "Accident Prevention Principles and Responsibilities," Bethlehem Steel Corporation, Bethlehem, PA, 1972, pp.3-16.

Eninger, M. V. "The Nature, Causes, and Results of Accidents," Bethlehem Steel Corporation, Bethlehem, PA, 1972, pp.3-48.

Eninger, M. V. *Job Safety Analysis*, Westinghouse Corporation, Pittsburgh, PA, April 1967, pp.1-21.

Eninger, M. V. "Developing Safe Job Procedures," Bethlehem Steel Corporation, Bethlehem, PA, 1972, pp.3-24.

Eninger, M. V. "Preventing Basic Types of Accidents," Bethlehem Steel Corporation, 1972, pp.3-32.

Grimaldi, John V. Simonds, Rollen H., *Safety Management*, Richard D Irwin Publishers, Homewood, IL, Third Edition, 1975, pp.459-460.

Haddad, Charles. "OSHA's New Regs Will Ease The Pain-For Everybody," *Business Week*, December 4, 2000, pp.90, 94.

Harris, Sydney. "Learning Requires More Than Repeating Bad Habits," *Chicago Sun-Times*, May 15, 1989.

Hendrick, K. M. "Working With JSA," *The Lifesaver*, Winter, 1980, Vol. 1.

Jackson, Jerry. "Accident deaths up, safety experts say," *Orlando Sentinel*, October 17, 2000, p.A9.

Johnson, William G. *MORT Safety Assurance Systems*, National Safety Council, Chicago, IL, 1980, pp. 193, 194, 254, 277, 292, 293, 294, 327, 501.

Journal of American Insurance. "An Introduction to Job Safety Analysis," July 27, 1981.

Kapp, Sue. "Why Job Safety Analyses Work," *Safety and Health*, April 1998, pp.54-58.

Karr, Al. "OSHA's Top 10 Violations," *Safety and Health*, December 2000, pp.34-38.

Kedjidjian, Catherine. "Speak No Evil," *Safety and Health*, September 1997, pp.38-39.

Krueger, Alan B. "Fewer workplace injuries and illnesses are adding to economic strength," *The New York Times*, September 14, 2000, p.C2.

Le Noble, Janet. *Workplace Safety in Action: Job Safety Analysis*, JJ Keller and Associates, Neenah, WI, 1996.

Liberting, G. Z. "Product Safety Hazard Analysis," The American Society of Mechanical Engineers, May 1977, Chicago, IL.

Manuele, Fred. *On The Practice of Safety*, Van Norstrand Reinhold Publishers, New York, 1993, pp.13, 37, 56.

Manuele, Fred. "Task Analysis for Productivity, Cost Efficiency, Safety and Quality," *Professional Safety*, April 2000, pp. 18-22.

National Safety Council. *Accident Prevention Manual, Eleventh Edition*, 1996, Itasca, IL, pp.130-139.

National Safety Council. *14 Elements of a Successful Safety and Health Program*, Chicago, IL, 1994.

National Safety Council. *Injury Facts, 2000 Edition*, Itasca, IL, pp.44-70.

National Safety Council. *Safety Agenda For The Nation*, Itasca, IL, 19 pages.

Petersen, Dan. *Safety Management: A Human Approach*, Aloray Publishers, Englewood, NJ, 1975, pp.256-257.

Petersen, Dan, Goodale, Jerry. *Readings in Industrial Accident Prevention*, McGraw Hill, 1980, pp.118-127.

Perkinson, Larry. "JSA: A New Look for an Old Friend," National Safety Management Society, *Occupational Hazards*, August 1995, pp. 63-65.

Roughton, James. "Job Hazard Analysis," *Occupational Health and Safety Canada*, Vol. 12, No. 1, January/February 1996, pp.41-44.

Roughton, Jim. "Managing a Safety Program Through Job Safety Analysis," *Professional Safety*, Vol. 37, No. 1, January 1992, PP28-31.

Roughton, James, Florczak, Cliff. "Make JSA's Work For You," *Safety and Health*, January 1999, pp.72-75.

Safety and Health. "OSHA Update-OSHA's Programmed Inspection Pace Jumps in Fiscal 2000," National Safety Council, December 2000, p. 16.

Shultz, J.D. "MSHA Hits Pay Dirt With JSA," *Safety and Health*, May 1991, pp.38-41.

Slote, Lawrence. *Handbook of Occupational Safety and Health*, John Wiley and Son, New York, 1987, pp.120, 121, 497, 498, 565-467.

Small Business Report. "Establishing A Safety Program," April 1978, pp.11-13.

Smith, L. C. "Let's Wed JIT and JSA, " *National Safety News*, January 1970, pp.75-77.

Spence, Stan F. "Case for Job Safety Analysis," *The Locomotive*, Vol. 61, No. 3, 1978, pp.55-60.

Stearns, Fred. "Job Safety Analysis Gives Steps to Uncover Accident Causes," *National Safety Council Power Press and Forging Newsletter*, March-April 1984.

Sutcliff, Virginia. "Improve Safety Through Job Safety Analysis," *Occupational Hazards*, October 2000, pp.55-56.

Swartz, George. "Effective Safety Programming," *Warehouse Safety*, Government Institutes Publishers, Rockville, MD, 1999, pp.11-34.

Swartz, George. "The Voluntary Protection Program," *Safety Culture and Effective Safety Management*, Itasca, IL, National Safety Council, pp.163-202.

Today's Supervisor. "Is Part of the Job Accident Prevention?" National Safety Council, October 1985, pp.14-15.

U.S. Department of Labor. "Safety and Health Program Guidelines; Issuance of Voluntary Guidelines" – 54: 3914 - 3916, 1/26/1989.

U.S. Department of Labor. Bureau of Labor Statistics, National Census Of Fatal Occupational Injuries, 1999, USDL 00-236, Thursday, August 17,2000.

United States Steel. "Principles of Accident Prevention in United States Steel," United States Steel Corporation, Pittsburgh, PA, Second Edition, 1965.

Vogel, Christine. "Cut Your Losses with JSA," *Safety and Health*, October 1991, pp.38-41.

Ziegle, John. "How to Keep the Dust off of Job Safety analysis," *Industrial Safety and Hygiene News*, October 2000, pp.65-66.

Index